Nicola Armaroli,
Vincenzo Balzani,
and Nick Serpone

Powering Planet Earth

Related Titles

García-Martínez, J. (ed.)

Nanotechnology for the Energy Challenge

2013
ISBN: 978-3-527-33380-6

Horikoshi, S., Serpone, N. (eds.)

Microwaves in Nanoparticle Synthesis

Fundamentals and Applications

2012
ISBN: 978-3-527-33197-0

Armaroli, N., Balzani, V.

Energy for a Sustainable World

From the Oil Age to a Sun-Powered Future

2011
ISBN: 978-3-527-32540-5

Olah, G. A., Goeppert, A., Prakash, G. K. S.

Beyond Oil and Gas: The Methanol Economy

2010
ISBN: 978-3-527-32422-4

Cocks, F. H.

Energy Demand and Climate Change

Issues and Resolutions

2009
ISBN: 978-3-527-32446-0

Wengenmayr, R., Bührke, T. (eds.)

Renewable Energy

Sustainable Energy Concepts for the Future

2008
ISBN: 978-3-527-40804-7

Nicola Armaroli, Vincenzo Balzani, and Nick Serpone

Powering Planet Earth

Energy Solutions for the Future

WILEY-VCH Verlag GmbH & Co. KGaA

The Authors

Dr. Nicola Armaroli
National Research Council (CNR)
Institute for Organic Synthesis
and Photoreactivity (ISOF)
Via Gobetti 101
40129 Bologna
Italy

Prof. Vincenzo Balzani
University of Bologna
Department of Chemistry
Giacomo Ciamician
Via Selmi 2
40126 Bologna
Italy

Prof. Nick Serpone
University of Pavia
Department of Chemistry
Via Taramelli 10
27100 Pavia
Italy

Library of Congress Card No.: applied for

British Library Cataloguing-in-Publication Data
A catalogue record for this book is available from the British Library.

Bibliographic information published by the Deutsche Nationalbibliothek
The Deutsche Nationalbibliothek lists this publication in the Deutsche Nationalbibliografie; detailed bibliographic data are available on the Internet at <http://dnb.d-nb.de>.

© 2013 Wiley-VCH Verlag & Co. KGaA, Boschstr. 12, 69469 Weinheim, Germany

Print ISBN: 978-3-527-33409-4
ePDF ISBN: 978-3-527-66742-0
ePub ISBN: 978-3-527-66741-3
mobi ISBN: 978-3-527-66740-6
oBook ISBN: 978-3-527-66739-0

Cover-Design Simone Benjamin, McLeese Lake BC, Canada
Typesetting Toppan Best-set Premedia Limited, Hong Kong
Printing and Binding Markono Print Media Pte Ltd, Singapore

To Claudia, Carla, and Linda

If you think education is costly try ignorance.

Derek Bok

Contents

Preface

This book originates from an Italian book published in 2008 (N. Armaroli, V. Balzani, *Energia per l'Astronave Terra*, Zanichelli Editore). It was an outstanding success in Italy, much appreciated by the public and by the scientific community, so much so that in 2009 it was awarded the top Italian prize for the dissemination of science (the Galileo Prize).

Since the interest in energy issues had witnessed a resurgence, particularly after the Fukushima accident, a second edition of the book was published in 2011 with updated relevant data and discussion on the consequences of the decisions taken at the Italian and European levels.

The success story of the book has led us to extend the energy issue to the scenarios of Canada, the United States and the United Kingdom. Some of the chapters of the original Italian version have been updated and all the data have been updated to mid-2012.

The book is written in a familiar style to reach the non-experts in energy matters and the general public. Nonetheless, we feel the messages embodied in this book would also be of interest to senior High School students and to first-year College students of Natural Science, Political Science, and Sociology, among others.

The young generation must be made aware that they will have to bear the brunt of our non-rosy energy legacy, and they therefore need to be informed of the energy issues facing Society in terms of the limited resources available on our spaceship Earth and our indiscriminate consumption of energy.

Only a collective effort will make the difference. We can't continue to use indiscriminately the limited resources of Nature. If we do, there will be consequences!

August 2012

Nicola Armaroli
Vincenzo Balzani
Nick Serpone

Introduction

> *Earth provides enough to satisfy every man's need, but not every man's greed.*
>
> Mohandas K. Gandhi

> *Nature provides a free lunch, but only if we control our appetites.*
>
> William Ruckelshaus

If you have a pair of shoes that need a new heel or a new sole, what do you do? That depends on how much you paid for them doesn't it? If you paid $599.99 when new, then it may be worth your while to have them fixed, but if you're part of the middle class (like us) and paid only $79.99 you'll likely throw them in the garbage as the cost of repairs may be too high relative to the original price. So you drive down to the mall and simply buy a new pair of your choice among the hundreds of different styles. Thank goodness you can still buy whichever pair you like and feel most comfortable wearing.

In one of his poems, Erri De Luca stated *I consider it a value saving water, repairing a pair of shoes . . .* Even if you wanted to repair your shoes you may find it difficult to find a cobbler today that would fix them. What's wrong with throwing out an old pair of shoes? You might mutter. Isn't our Society one of consumerism? Isn't our civilization one of use-and-discard?

The Italian philosopher Umberto Galimberti wrote *consumerism is the first of the seven deadly sins of our age*–a model *that provides a false sense of welfare based on the destruction of resources and on the exploitation of people.*

What if, instead of being blinded by the glint of consumerism, you try to understand how things really stand in the global reality? If you did, you might discover that the new pair of shoes is manufactured with materials obtained from fossil fuels (plastics, adhesives, paint), and that the energy consumed to produce that pair of shoes (electrical, mechanical, thermal, and luminous) was also produced from fossil fuels. You might even discover that the manufacturing of that pair of shoes produced various waste substances, to which you have now added your old shoes. And not least, you might be surprised to find out that the fine new pair of shoes you just bought was produced by workers–maybe child labor–poorly paid and poorly protected in some sweat shops in some country where environmental laws are lax or non-existent and where pollution is too often at unsustainable levels.

Powering Planet Earth: Energy Solutions for the Future, First Edition. Nicola Armaroli, Vincenzo Balzani, and Nick Serpone.
© 2013 Wiley-VCH Verlag GmbH & Co. KGaA. Published 2013 by Wiley-VCH Verlag GmbH & Co. KGaA.

In buying the new shoes you probably thought you had not been influenced by incessant TV ads that professed that this or that pair would make your walks more comfortable, better for your back, and so forth. . . . . Of course, the new pair of shoes came in a cardboard box filled with wrapping paper (produced from trees) that you brought home in a plastic bag also produced from fossil fuels. Once you're back home, you'll throw away the box, the paper and the plastic bag into the garbage bin, possibly paying no attention to sorting these wastes for possible recycling.

This discussion on the pair of shoes also applies every time you buy electronic products such as a computer, a mobile phone, a TV set, and even food at the supermarket. These products will create an even more serious problem of resource consumption and generation of wastes. For instance, some 500 Italian supermarkets throw away annually around 55 000 tons of food in the garbage, even though the expiry date had not been reached – food that could still be safely eaten.

Our consumer model is based on a crazy vicious circle of production-consumption-production. Goods are produced to meet demand. At the same time, also produced are the needs to ensure a continual manufacturing of goods. These goods must be consumed quickly so they can be replaced. They mustn't be too fragile, however, otherwise no one would buy them. It is sufficient that only one part be fragile. But *spare parts* either do not exist or they are sold at such a high price that makes it inconvenient to repair them.

Even if there were no need to replace a product, that *need* would then be instilled into the consumer, albeit very subtly, by means of unceasing media advertising to persuade you to discard used goods to make way for new ones. In this game, fashion often takes over where advertising fails to reach you – the consumer. This is yet another subtle strategy to overcome the consumer's resistance to throwing things out. In fact, fashion makes it socially unacceptable to continue wearing a suit or a dress that is still perfectly usable.

Whenever resources are used to produce a product (for example, a car), or to operate a service (for example, heating a swimming pool), wastes are inevitably generated that often take up more space than the resources did. Depending on their state, wastes tend to accumulate as *lumps* on the surface of the Earth – later they will find their way to surface waters or even the aquifers, often a long way from the waste source. Gaseous wastes will wind up in the atmosphere. You are no doubt aware by now that waste treatments have become increasingly very complex problems.

By definition, at the end of any industrial process that produces goods and services, the natural environment becomes depleted of its original matter only to be **replaced** – an ugly word, but better than **enriched** – with quantities of solids, liquids, or gaseous wastes. Waste substances change the nature of the soil, the water and the atmosphere, often making these three ecosystems unsuitable to perform their vital functions. In the final analysis, you mustn't forget that the economy is ultimately sustained by natural resources.

It is unfortunate that every call to decrease consumption, starting with pleas for energy conservation, is in plain contrast with today's dominant notion – supported by several economists and embraced by most politicians – according to whom it is necessary that a country's Gross Domestic Product (GDP) increase by at least 2–3% annually. Don't forget, however, that an increase in GDP also implies an increase in resource consumption and waste production. The Second Principle of Thermodynamics (we'll have more to say on this later) tells us that it is impossible to create a perpetual motion machine. Likewise, it is impossible to continue with our voracity to consume resources as though they were infinitely available for the picking. They are definitely not! They are finite! This is a reality with which economists and politicians, and most importantly consumers, must come to terms.

At this point you're asking yourself: *I just wanted to read a book on energy, so what's with all this stuff*–resources, conservation, consumerism, thermodynamics. Don't worry; you're not going to be left in the cold. In the rest of the book, we'll tell you about energy–energy from chemical bonds to the law of Einstein, energy from coal to nuclear energy, energy from food (obesity) to solar energy.

With this introduction we wanted you to be aware–at an early stage–that energy is a crucial resource that needs to be understood in the kind of world we live in. So far, the great accessibility to energy resources has made you experience a life immensely more convenient and enjoyable than your grandparents did. At the same time, you must realize that your lifestyle can further the degradation of the planet and seriously affect the quality of life, especially that of the next generations and most particularly that of your grandchildren and great-grandchildren.

The greatest challenge and opportunity that mankind is faced with is to alleviate the main problem that plagues it. Mankind needs to develop new sustainable energy sources and related technologies. It needs to know the fundamental laws that concern energy. Everybody needs to be informed on current energy systems, and needs to have some ideas on the prospects of new technologies that can assist people become more aware, better informed, and more responsible.

In short, after you've read this book, the next time your pair of shoes needs new heels or new soles you might decide to have them fixed–you may even try to fix them yourself. If you do either, then we've done our job!

1
What Is Energy?

> *It is important to remember that we are energy.*
> *Einstein taught us that. Energy can neither be created nor destroyed; it just*
> *changes form.*
>
> Rhonda Byrne

You're reading a book. Close your eyes for a moment and remain perfectly still. Under these conditions, you might perhaps think that you're not consuming energy. Well, not quite. For when you breathe, your brain works, your heart throbs, and your body has a different temperature (probably higher) than its surroundings. All this costs energy–energy taken from what you've eaten at this morning's breakfast or at dinner last night, or else energy drawn from the fat reserves that have accumulated in your belly, your hips, or some other parts of your body.

At some point during the week you'll likely participate in some form of sport activities (jogging, swimming . . .), and you'll probably experience a feeling of great well-being. The effort made in such activities stimulates the release of endorphins and neurotransmitters, which induce pleasure. However, after a nice swim your energy content will be lower than before. Don't believe for a minute that the so-called *energizing shower foam* will recharge your battery. In fact, it would be better to have a snack somewhere. If you drive and have to stop to fill-up at the gas station, you will likely complain about the latest fuel price increases. And if you're thirsty, you're likely to buy a bottle of water or a bottle of pop at the seven-eleven–have you noticed that a liter of bottled water costs more than a liter of gasoline? And to think that over 60% of the price of fuel represents indirect taxes (excise taxes, sales taxes, etc.–at least in Europe) that all go to the Treasury (in the case of water, the government takes in only 4–5% . . .). Unfortunately, we seldom pay attention to these hidden taxes, and so we tend not to complain–would it change anything if we did?

Once home, back from a hard day's work, it may be time for a well-deserved snack: perhaps a banana or a kiwi. If you do snack, look at the stickers to see where these fruits came from. You discover that the banana came from Costa Rica, the kiwi from New Zealand. So to reach your table these fruits had to travel some thousands of miles. You eat them with gusto and you feel much better. Next you'll turn on your personal computer to check your e-mail or access the social networks, or otherwise surf the web.

Powering Planet Earth: Energy Solutions for the Future, First Edition. Nicola Armaroli, Vincenzo Balzani, and Nick Serpone.
© 2013 Wiley-VCH Verlag GmbH & Co. KGaA. Published 2013 by Wiley-VCH Verlag GmbH & Co. KGaA.

You can't complain. It's not been a bad day, for in a short time you've achieved much: maybe you read a book, you went for a swim, you went for a drive, you had a snack, or maybe you just chilled out doing nothing at home. You probably don't realize it, but all this was made possible thanks to an enormous availability of energy: for instance, the energy of the cells in your body, the energy from the boiler, the energy from the car's gas tank, the energy of the ship that sailed the oceans to bring you the banana and the kiwi, and not least the electrical energy from the utility network.

If now you asked yourself: what is energy? You'll probably have no idea of how to define this omnipresent entity in your life in clear and concise terms. In fact, it may even prove embarrassing, because we usually like to know only what's around us and tend to be suspicious of that which we don't know and can't see. Don't be too distressed: energy ignorance is widespread, and understandably so. Energy is an elusive concept and only seemingly intuitive. It is so difficult to define that for millennia even scholars gave vague definitions or even completely wrong ones. For instance, the 7th Edition of the *Encyclopedia Britannica* of 1842 defined energy as *"the power, virtue, and efficacy of a thing."*

If we've come to understand the notions of what energy is and what laws and principles govern it, it is mostly thanks to the passionate and prolific insights of a small group of curious men that, since the end of the eighteenth century, dedicated much of their time to this problem: men such as James Watt, Sadie Carnot, Justus von Liebig, James Joule, Rudolf Clausius, William Thompson (better known as Lord Kelvin), Ludwig Boltzmann, Walther Nernst, and Albert Einstein.

Energy and Related Terms

The concept of *energy* is not immediately definable. Before we attempt to understand what energy is, we need to define another concept that precedes it: *work*.

Work can be described as the use of a force to move something. The amount of work depends on how much force is used and the distance the object is moved to. From a mathematical point of view, *work* is the product of *force × distance*.

We do work when we lift a weight against the force of gravity, such as, for example, lifting a crate of apples. The magnitude of the work needed depends on the mass being moved (how many apples are there in the crate?), the magnitude of the gravitational force (whether we're on Earth or on the Moon) and the height to which we want to lift the object to: on the table? – on the shelf above?

Often the mass may be that of our bodies: for example, we do work when we climb the stairs or a ladder. Since the force of gravity is identical in the Italian regions of Valle d'Aosta and Abruzzo, and the mass to be moved is constant over the years (provided we maintained our figure), greater work will be needed to climb to the top of Mont Blanc, 4810 meters, than to climb to the top of the Gran Sasso at 2912 meters in the Apennine mountain chain of Italy.

If you attempted to move an object (for example a 4-wheel drive SUV with your arms) and were unsuccessful, then you've done no work. In common parlance, however, *work* can mean other things. For instance, a letter carrier and a notary both do *work*. However, from the scientific point of view, the carrier does more *work* than the notary, although you would not intuitively think so from their standard of living. But this has nothing to do with science.

How then, would you describe the ability of a system (for example, a liter of gasoline, a living being, a rock that falls, a car . . .) to do work? What is the parameter that quantifies this ability to do work? We're getting there: the ability to perform work is *energy*, not to be confused with *power*, which describes the rate at which *energy* is transferred, used, or transformed. In other words, *power* refers to the mathematical relationship between energy and time: *power = energy/time*).

Consider, for example, two athletes with the same body mass that compete in the 100-meter final at the Olympic Games. They both do exactly the same work in this glorious event; the one that uses up even an iota of more power will reach the finish line first. That greater effort or work will suffice to make the difference between an Olympic medal and total oblivion.

From One Energy Form to Another

At this point we can go a little further and free ourselves from the concept of work being purely mechanical, although it may sometimes be just that (the crate of apples). Thus, any process that *produces a change* (maybe the temperature, the chemical composition, speed, or position) *in a certain system* (a living organism, an inanimate object, a car) is deemed to be *work*.

Broadly speaking, the ability to do work manifests itself in many ways; what we define as *forms of energy* go far beyond muscle energy described above. In their diversity, all forms of energy have one common feature: they are always the expression of a system that is capable of exerting a force, which can act against another force.

We can easily locate seven forms of energy, almost all of which we experience daily:

1) **Thermal energy:** radiators that heat our house.

2) **Chemical energy:** natural gas that feeds our gas furnace and/or gas stove.

3) **Electrical energy:** energy that makes electrical appliances work.

4) **Electromagnetic energy or light:** sunlight that makes plants grow in a vase, on a balcony, or on a farm.

5) **Kinetic energy:** energy of a glass bowl falling to the ground (gets broken).

6) **Gravitational energy:** If the glass vase falls from a height of 10 centimeters (about 4 inches) it will likely not break, but if it falls from 2 meters (about 6.5 feet) there's no hope of saving it.

7) **Nuclear energy:** energy from the atom: difficult to see – we'll have more to say on this later.

Table 1 Different forms of energy and various methods with which one energy form can be converted into another energy form.

From \ To	Thermal	Chemical	Electrical	Electromagnetic (light)	Kinetic
Thermal	–	Endothermic reactions	Thermo-ionic processes	Lamps (tungsten wired)	Motor engines
Chemical	Combustion		Batteries	Fireflies	Muscles
Electrical	Electrical resistances	Electrolysis	–	Electro-luminescence	Electric motors
Electromagnetic (light)	Solar collectors	Photosynthesis (chlorophyll)	Photovoltaic panels	–	Solar sails
Kinetic	Friction	Radiolytic reactions	Electrical alternators	Accelerated charges	–
Nuclear	Fission and Fusion	Ionization	Nuclear batteries	Nuclear weapons	Radioactivity

Note: none of the energies in the first column can be transformed into nuclear energy.

The various forms of energy can then be converted from one form to another, but not always. For example, we can transform the Sun's light energy into electricity through a solar panel. However, contrary to what is often thought, we cannot transform nuclear energy directly into electrical energy. Nuclear power plants are nothing more than sophisticated water kettles that convert nuclear energy into thermal energy, which in turn is converted into mechanical energy and then finally into electrical energy.

If you wish to have other examples of energy transformation, think of your typical day and unleash your fancy; you may find some inspiration in Table 1.

Sources of Energy

Energy *sources* are physical entities from which it is possible to obtain one or more *forms* of energy. These sources may be very different:

1) **Plant and mineral resources:** in the case of coal, oil, gas, and biomass, the chemical energy is stored in carbon-carbon (C–C) and carbon-hydrogen (C–H) chemical bonds; to free this energy requires a trigger and an oxidizer (oxygen); in the case of uranium, the energy is of the nuclear type and can only be freed by fragmentation (fission) of the atomic nucleus.

2) **Artifacts:** If a river were blocked by a dam, it would be possible to transform the gravitational potential energy of water into kinetic, mechanical and electrical energy through a series of pipelines and machinery; similarly, wind turbines can convert the kinetic energy of moving air mass.

3) **Celestial bodies:** the Sun is a source of light energy; the Earth is a source of thermal energy (underground) and gravitational energy (the pot that falls).

It's good to remember that energy sources are not sources of energy only – they can also be sources of some useful products. For example, with fossil fuels we can manufacture a variety of useful plastics, fertilizers, and medicines (among others). With a dam, we can control the flow of water in a river; as for the Earth, we need not emphasize that it is useful for many other purposes.

Energy sources are said to be *primary* sources if they are directly available in nature – for example, fossil fuels, sunlight, wind, moving water (as in rivers), vegetation, and uranium. These can be used as such or can be converted into other forms that are referred to as *secondary* energy sources; these are more easily used: for example, products derived from crude oil (fossil fuels in general).

The forms of energy – whether primary or secondary – typically used are referred to as *final forms*; among these are electricity and gasoline. By contrast, neither solar radiation nor crude oil belongs to this group – the latter needs to be refined before use.

The Pillars of the Universe

The first scientific and experimental studies on the transformations of energy date back more than two centuries when machines were used to transform heat into motion, and *vice versa*. Historically (and logically), this branch of physics became known as *Thermodynamics*.

In the nineteenth century, men who laid the foundations of thermodynamics during the years of great technological advancement were mostly British, French and German. They were often driven by the desire to contribute to the development and technological supremacy of their country.

Thermodynamic studies conducted in the second half of the 1800s led to the formulation of some basic laws, or principles, whose validity can be extended to all forms of energy. In other words, without realizing it, the thermodynamicists of that era went beyond their original ambition. They wanted to understand the operation of simple machines and in doing so managed to uncover some of the fundamental pillars that hold the universe together.

The two principles of thermodynamics are so basic that often they are referred to simply as the First and Second Principle of Thermodynamics. Incidentally, the capital letters are not typographical errors. Before illustrating these Principles, it is useful to clarify briefly some of the concepts underlying these Principles, namely *temperature* and *heat*.

Particles in Motion

Thermal energy (or *heat*) is a manifestation of the ceaseless movement with which atoms are agitated – atoms are the submicroscopic particles that make up matter. As for *temperature*, we are all convinced that we know what it is: who has never used a thermometer? However, the concept of temperature is far less trivial than it seems at first. It is rigorously described according to the average kinetic energy of motion of the atoms.

Here we shall limit ourselves to state simply that temperature is a property that defines the direction of the transfer of thermal energy from one system to another. Thermal energy (heat) tends to move from a system of *higher temperature* to a system of *lower temperature*. The process stops when the so-called *thermal equilibrium* is reached, at which point there is no longer transfer of heat energy between the two bodies (macroscopically speaking) since they are at the same temperature.

The *scale* used to measure temperatures is based on a simple convention. You can use whichever scale you like (Celsius, Fahrenheit, Kelvin). Don't be surprised, then, if you find yourself in the United States during a snowstorm and are told that the outside temperature is 32 degrees (Fahrenheit, °F), equivalent to 0 °C (or 273 °K).

Heat (Warmth) – an Exchangeable Energy

Heat is thermal energy that can be exchanged between two bodies of different temperatures. For millennia, it was believed that heat was an intangible fluid (maybe someone still believes it . . .) – but this is not true. When water is heated in a pot, the flame does not directly heat the water but warms the bottom of the pot, which in turn heats the water. This is an example where exchange of heat takes place between three bodies (from the flame, to the pot, to the water).

Atoms and molecules that constitute the flame (which technically speaking is called *plasma*, a very hot form of ionized gas) move, rotate, and vibrate rapidly. These particles collide with the bottom of the pot and stimulate the vibration of atoms of the metal (not their change of position, at least as long as the pot does not melt . . .). This chain transfer process proceeds rapidly until it involves the water molecules inside the pot, starting from the first layer in direct contact with the metal.

If we keep the flame lit, the water will come to a quick boil, and only then can we throw in the pasta. But if the bottom of the pot were perfectly insulated, we would have to resign ourselves to eating uncooked pasta or else starve, as the water will remain cold forever.

You Can't Run Away from Them – the Principles of Thermodynamics

The *First Principle* states that the energy of an isolated system, that is a system that cannot exchange matter or energy with its surroundings, is always the same; it can convert energy from one form to another, but the total amount remains unchanged. Thus the energy of an isolated system – for example, the universe – is always constant.

Objectively, the first principle is good news, though a bit distressing for those who wished to stay on a diet: the energy of the food eaten is either spent through mental or physical exercises, or else it accumulates as fat in various parts of your body (belly, hips, . . .).

The chemical energy stored in the gas tank of a car will take us to some vacation spot by doing work, and so we might believe unknowingly that the engine has literally "eaten" all the energy available in the gas tank. Well it's not really so. If we managed to get to the Stelvio pass (2760 m, Italy), for example, the chemical energy stored in the fuel purchased at the gas station was converted inside the engine in a process involving air—in part—into gravitational potential energy (we and the car are now at a greater height than before), in part as heat emitted by the car exhaust, and in part in the form of friction between the tires and the road.

The mass of fuel was converted to gases, mostly water vapor (H_2O) and carbon dioxide (CO_2), that were discharged into the atmosphere. In this transformation, the initial volume of the fuel increased some 2000 times because the gases produced are much less dense. But since the gas is invisible, we have no guilt feelings of having polluted the air we breathe. We no longer see anything, but energy isn't lost. The unobtrusiveness with which the fuel disappeared is truly amazing.

The *Second Principle* is one of nature's most fascinating laws. The resulting consequences are vast. They can be formulated in various ways, but the most intuitive is probably the following: in an isolated system, thermal energy is always transferred from a body of higher temperature to one of lower temperature.

It's important to point out that the Second Law doesn't say that heat cannot pass from one cold body to a warm one. The way the refrigerator works is precisely for this reason, and there is no doubt that it functions. But the refrigerator is not an isolated system. The Second Principle states that if we want heat to flow in the direction opposite to its natural tendency, then we need to provide power to the system: the refrigerator works only if it's connected to an electrical power outlet.

The *Second Principle* leads us very subtly to the notion that there exists a *hierarchy* between the various forms of energy. Note that every time you do some form of work, you consume energy; the resulting heat is dissipated to the surroundings.

Thermal energy will make its presence felt in any process that involves energy conversion. For example: the car engine and the motor of the refrigerator get hot; our body is warm; without cooling towers, the nuclear power station would undergo a meltdown.

All forms of energy can be transformed completely into heat, that is, thermal energy; the opposite process cannot and does not happen. Every time you convert a noble form of energy into another, for example, electrical energy into mechanical energy, not all the available quantity can be used to accomplish useful work. Inevitably, a part will be degraded into thermal energy forever.

In most cases, this thermal tax is characterized by the thermal environment, primarily the atmosphere and surface waters. This explains why power stations are built near the seashores, near lakes, or near rivers. Even though a power plant is built solidly, it cannot directly convert even half of the fuel's chemical energy into electricity. Most of that energy turns into heat, which is discarded in the immediate vicinity of the power station. Even nuclear power plants have an output that does not exceed 30–35%: only about a third of the heat generated in the reactor is converted to electricity, while the remaining two thirds is relinquished to the environment by the cooling towers, and so is lost. For comparison, a thermo-electrical gas-fed power station that uses combined cycles can reach a yield close to 60% – that is, nearly two thirds of the energy is converted to electricity.

It is unfortunate that no ship sailing on a river can operate its engines using the heat dissipated by the numerous power stations situated on its banks. The reason is that the heat dissipated by the power plants has a much *lower value* than the chemical energy of the fuel. Hence, its exploitation to useful purposes is rather limited. The same applies to a car. A good part of the compact and valuable energy initially stored in the gas tank will be dispersed in a myriad of unnecessary forms of heat – for example, friction, already mentioned earlier. In these processes, the energy of the universe is nonetheless preserved, in keeping with the First Principle, but loses value to comply with the Second Principle. Whoever is still convinced that he can build a perpetual motion machine knows perhaps the First Principle, but obviously ignores the Second Principle.

In more general terms, the Second Principle tells us that a profound asymmetry exists in nature: disorder is obtained in an instant, while to restore order from chaos necessitates time and effort.

Inherently, natural systems tend spontaneously toward disorder. The universe is made this way. Hence, we need to find an explanation as to the reason why this is. The spontaneous and inexorable trend that energy is transformed into its most disorderly form – heat – is one of the many expressions of the general tendency of the universe toward chaos. This is expressed scientifically through a function we call *Entropy*. Though the energy of the universe is constant, the entropy increases. To illustrate this concept, imagine putting a layer of 100 red marbles in a box, then overlay this layer with a layer of 100 blue marbles and then again a layer of 100 green marbles. If we now shake the box vigorously, the marbles will mix. Ultimately a state will be reached at which even if we continue to shake the box for millions of years, it is highly unlikely (in fact impossible) that we will regain the original orderly configuration.

A small reflection tells us that our daily life is a continuous demonstration of the implacable power of the Second Principle: to mess up our room requires but a minute (and a little effort), but to put it back in order, it takes hours of hard work. At this point you might be tempted to think that living beings do not obey the Second Principle. Unfortunately, this is merely an illusion. The tendency toward disorder (entropy) should be measured in relation to the environment that surrounds a given system.

Order represents the extraordinary complexity of all forms of life (even the simplest ones), that are largely balanced by the disorder generated from the progressive consumption of the Sun's energy, from which we are not isolated. But it's not all. For living beings to survive – that is, to remain in an ordered state – they continually produce wastes (a form of disorder) that are discharged into the environment, starting with those physiological ones (pardon the expression – going to the toilet).

The First and Second Principles should be a basic part of the cultural preparation of each of us, just like the alphabet, multiplication tables, the Constitution, and *The*

Divine Comedy. Unfortunately this is not the case. Every day we hear journalists mention that incinerators destroy wastes and produce energy. Economists and union leaders are confident that economic growth has no limits. Environment ministers talk about clean coal. Some scientists deny global warming. Maybe their refrigerator works without being connected to an electrical outlet.

Einstein's Equation: E = mc²

This equation is well known. It is the icon of the twentieth century. It's sometimes seen on T-shirts just as are the names of a pop group or a photo of Che Guevara. This equation defines energy in such a way that anyone can understand it, even if (in fact) it's a little difficult to accept.

$E = mc^2$ means that mass and energy are the same thing albeit under different guises. As the ice melts it turns into water, totally changing its appearance, so is mass a form of *frozen* energy that can be converted into more familiar forms: kinetic energy, thermal energy, and so on.

In the formula, the letter c represents the speed of light in vacuum. Raised to the second power, it has an even larger numerical value. So, since the right and the left hand sides of the Einstein equation must be numerically equal (otherwise, what kind of equation would it be?), and since c^2 is on the side of m, to obtain massive amounts of energy we need only convert small quantities of mass.

Every time you produce energy of any kind, quantities of mass – large or small – largely disappear. This *dematerialization* recalls some improbable science fiction movies and makes us a little bit skeptical. But that's the way it is, folks. The energy consumed in a month from a huge megalopolis – for example, modern London – is comparable to the energy *frozen* in the mass of this book. The unfortunate destruction of Hiroshima and Nagasaki in World War 2 occurred by converting only a few grams of matter into energy; a small amount, but certainly a measurable one.

Nuclear fission allows the conversion of materials into energy very efficiently, but, as we shall see later, it leaves extremely hazardous wastes. A kilogram of uranium in a nuclear power plant can generate 50 000 kilowatt-hours of energy, while 1 kg of coal in a thermal power station produces only 3 kilowatt-hours. Einstein's equation is valid in both cases. The amount of matter that *evaporates* to become energy is dramatically higher in uranium than in coal.

For nearly 5 billion years, the Sun has converted 4.4 billion tonnes of hydrogen every second into electromagnetic energy through nuclear fusion processes at temperatures well above 10 million degrees. A tiny fraction of this endless energy flux lightens our days. Of course, Einstein's equation also suggests that it is possible to convert energy into mass. This has been verified by means of some very complicated experiments. It is possible to create new particles of matter by concentrating huge amounts of energy into a small volume of space.

From Kilowatt-hour to the Barrel of Oil

Units of measurement are the despair of many High School, College, and University students. There are some units that are common and easily understood by all. Others are more difficult to digest. The so-called international system of units (SI) defines the unit of measurement of seven physical quantities: *length* is measured in meters (m), *time* in seconds (s), *mass* in kilograms (kg), *temperature* in degrees kelvin (K), *amount of a substance* in moles (mol), *electrical current* in amperes (A), and *light intensity* in candelas (cd).

All other physical quantities, strange as it may seem, are a combination of these seven units of measure. Some sadistic science teachers like to see students cringe when told that electrical resistance has something to do with kilograms, or that heat capacity has something to do with meters. Many students never understand this and forever drop their scientific studies.

As we had anticipated, energy is not a primary physical concept. It may seem bizarre that, from this point of view, electricity and light intensity are both hierarchically superior to energy, but that's the way it is, folks.

We have already stated that *work* can be expressed as the product of *force* multiplied by *distance* (length). In terms of the size of the physical parameters indicated in brackets we have:

$$[work] = [force] \times [length]$$

In turn, *force* is a parameter derived from the next equation; that is, it can be expressed as *mass* times *length* divided by *time* raised to the second power:[1]

$$[force] = [mass] \times [length] / [time]^2$$

Accordingly, *work* – that is, *energy*, which represents its quantification – has the following physical dimensions:

$$[work] = [energy] = [mass] \times [length]^2 / [time]^2$$

However, no one is thrilled to have to use a unit of measure as twisted as kg-m^2/s^2 to express a quantity of energy. Fortunately, new units have been adopted for sizes derived from these fundamental parameters, often indicated by the names of famous scientists of the past. For instance, in the case of energy, it was decided that the unit kg-m^2/s^2 could simply be called a *joule* and would be represented by the capital letter J.

By contrast, the watt (symbolized as W) is the unit of power: 1 watt equals 1 joule divided by 1 second (W = J/s). The choice of so honoring Joule and Watt was certainly appropriate, considering the contribution of these two British scientists to the advancement of knowledge in the field of energy.

Unfortunately, the joule is a very small unit of measure. A small field-mouse consumes about 50 000 J per day to survive. The gas tank of a medium-sized car

---

1) The famous Newton's law $F = ma$ shows that *force* equals *mass* times *acceleration*, which in turn is a change in *velocity* (defined as *length* divided by *time*) per unit time.

Table 2 Some energy units in common use.

Units	Symbol	Value in joules (J)
Calorie	cal	4.19
British thermal unit	BTU	1.05×10^3
Kilowatt-hour	kWh	3.60×10^6
Barrel of oil equivalent	boe	6.12×10^9
Tonne of oil equivalent	toe	4.19×10^{10}

Table 3 Symbols and prefixes of multiples and sub-multiples.

Symbol	Prefix	Factor	Symbol	Prefix	Factor
a	atto-	10^{-18}	k	kilo-	10^3
f	femto-	10^{-15}	M	mega-	10^6
p	pico-	10^{-12}	G	giga-	10^9
n	nano-	10^{-9}	T	tera-	10^{12}
μ	micro-	10^{-6}	P	peta-	10^{15}
m	milli-	10^{-3}	E	esa-	10^{18}

contains over one billion joules of energy. Hence, for convenience we use energy units of much greater magnitude. Among the most common are the *kilocalorie* used to measure heat and the *kilowatt-hour* to measure electrical energy.

Compilation of energy balances in the world often uses other measurement units which are not strictly related to the physical quantity of energy, as indicated in Table 2.

Also commonly used are units of mass or volume of fossil fuels, to which are associated a certain energy content. The most often used is *toe* (*tonne of oil equivalent*), which represents the heat developed by the complete combustion of one ton of oil; also used is its sub-multiple kilogram of oil equivalent (*kgoe*). The barrel of oil equivalent (*boe*) is also greatly used, which corresponds to the energy developed from the combustion of 159 liters of crude oil (approximately 130 kg).

The amount of energy involved in the large variety of natural and artificial processes can vary immensely. For example, for a flea to jump requires a one hundred millionth of a joule; a tropical hurricane develops an energy equal to tens of billions of billions of joules.

Thus, if we wish to maintain the same unit of measurement for whatever energy phenomenon, it would be better to use the conventional prefixes for multiples and sub-multiples shown in Table 3, so as to avoid the burden of many zero digits.

From a Chemical Bond to a Tsunami

Let us now take a short trip on the energy scale starting from two infinitesimal entities that can appear insignificant at first, but that in reality maintain the treasure of fossil fuel energy. We're referring to the chemical bonds between two carbon atoms (C–C) and between a carbon atom and a hydrogen atom (C–H). Each of these bonds contains about 0.7 billionths of billionths of a joule, that is, 0.7 attojoules (otherwise written as 0.7 aJ).

This is small change in the currency of events on which the industrial civilization, the digital age, and the globalization of the economy are based – in short, modernity. To get this money, which too often has literally dictated the price of the economic currency, there's been no hesitation to resort to war, unfortunately.

To hit a key on a computer's keyboard consumes 20 thousandths of a joule (20 mJ). A well-fed adult takes on an average 10 million joules (10 MJ) a day. A kilogram of good quality coal contains about 30 million joules of energy, that is, 30 megajoules (30 MJ).

The annual world consumption of primary energy today is around 510 billion billion joules, that is some 510 esajoules (510 EJ). Of these, four fifths, or about 410 EJ, are from fossil fuels. The largest hydrogen bomb tested so far has developed 240 million billion joules (240 PJ), an energy 3000 times greater than the bomb dropped on Hiroshima (84 trillion joules, 84 TJ).

Each year the Earth receives from the Sun 5.5 million billion billion joules (5 500 000 EJ) of light energy; approximately 2000 EJ are converted into new biomass through the process known as photosynthesis. At this time, it would also be interesting to describe briefly the power in some phenomena, that is, the amount of energy per unit time. For instance, a traditional incandescent bulb absorbs 60 W. A washing machine that works at 60 °C requires approximately 800 W. The engine of a Ferrari Formula 1 car can develop 550 000 watts (550 kW). The four engines of a transcontinental Boeing 747 jumbo jet produce 80 million watts (80 MW) on take-off. By comparison, a violent thunderstorm develops around 100 billion watts (100 GW).

The average quantity of energy consumed every second on a global scale amounts to about 16 trillion watts (15 TW), a value obtained by dividing the annual global energy consumed (510 EJ) by the number of seconds in a year (about 31.5 million). A volcanic eruption can disburse 100 trillion watts of power (100 TW). An earthquake of magnitude 8 on the Richter scale releases 1.6 million billion watts (1.6 PW) and can produce huge oceanic wave surges, thereby generating tsunamis that can bring death and destruction to the mainland. These numbers give you a rough idea of the immense power of nature and of the respect that nature, therefore, deserves from mankind.

2
Yesterday and Today

The struggle for existence is the struggle for power.

Ludwig Boltzmann

To better appreciate the energy problem, we need to consider the Earth as a giant spaceship traveling in an immense Universe at a speed of 29 kilometers per second (about 18 miles per second). The energy resources that spaceship Earth uses are not consumed just for its propulsion. Much of them are, in fact, consumed to carry the many passengers and crew. At the last count (November 2011), there were about 7 billion passengers, and this figure is likely to increase to about 8 billion within the next 20 years. Demographics tell us that the annual increase in the world's population is around 80 million units, taking place mainly in developing nations: for example, 37 Indian and 15 Chinese babies are born every minute.

All the inhabitants of the Earth aspire to greater material well-being. To achieve this objective, however, requires energy. With energy you can do anything, or almost anything. You can also remedy the shortage of other essential resources. For example, it is often said that drinking water, which in various regions of the world is beginning to run low, will be the oil of the twenty-first century. The aggravating circumstance is that while oil may be replaced by other sources of energy, fresh water has no substitutes except . . . the use of energy. Sea water can, in fact, be converted into drinking water, but at a price: about one liter of oil for every 3 cubic meters of water.

Over the past 150 years, for good or ill, our life has changed dramatically thanks to the ready availability of energy associated with the exploitation of fossil fuels. So let's briefly look at the most important aspects of this change.

The Energy Slaves

For several millennia, mankind has derived energy from the muscular work of men and animals, from wind power (windmills and boats), from waterways (river navigation; watermills) and from biomass (timber).

In the great civilizations of the past – Egyptian, Chinese, Greek, Roman – an important energy source consisted of slaves. Without them there would have been

Powering Planet Earth: Energy Solutions for the Future, First Edition. Nicola Armaroli, Vincenzo Balzani, and Nick Serpone.

Figure 1 The electrical power required to operate a television set is equivalent to the power developed by the work of two people; a washing machine would require the continuous work of some 15 people.

no pyramids, no Great Wall of China, and no Colosseum in Rome. Slaves were mostly prisoners of war; but could also have been debtors and convicts.

Slavery was widespread in the 1700s and 1800s, particularly in America's southern States, where for decades millions of Africans were imported and forced to work on farms, on plantations, or as domestic help. Although slavery has been abolished officially for some time, even today we see something not too different (child labor) in various parts of the world.

A man in good health can generate an output of about 800 watts (W) for a short period of time, as, for example, in running up a flight of stairs. However, in an ongoing activity that lasts many hours man will likely fail to develop a power level higher than approximately 50 W. Hence, we can estimate that for a 12-hour working day a slave will develop an amount of energy corresponding to about 600 Wh (watt-hours).

Let's now see how the energy produced by an *energy slave* compares with the energy consumed by the various appliances and gadgets we use daily. Take, for example, a radio-CD, which has a power requirement of about 25 W. This means that its operation consumes a quantity of energy that is around half that produced by the work of a slave. Watching a football game on a 30-inch LCD television set uses approximately 100 W of electric power, equal to that produced by two energy slaves (see Figure 1). The use of a personal computer requires a power of about 150 W, equivalent to the work of three slaves.

As illustrated in Figure 1, doing laundry with a Class A washing machine (i.e., one of the most efficient, which consumes about 800 Wh for a wash at 60 °C) is equivalent to using an hour's work of some 15 slaves.

A 10-minute use of a hairdryer (power: 1.2 kW) consumes 200 Wh, a quantity of energy equal to that produced by 4 energy slaves for about an hour's work. Heating ourselves with a small electric radiator (2.5 kW) is equivalent to using the energy generated by the work of 50 slaves.

A simple lawnmower (not the kind you sit on and ride) has a power of 3.5 kW. In an hour's work, this machine consumes an amount of energy equal to that produced by 6 slaves for a 12-hour period.

A mid-sized car engine, which produces an output of about 80 kW traveling at a cruising speed, does work equal to that of 1600 slaves. It is evident that even the Roman Emperor Caesar Augustus could not afford the luxury of instant availability of such a number of slaves by as simple a gesture as turning the key on the dashboard.

One of the most powerful means available today of transporting passengers is the Boeing 747-400. When fully loaded, this develops on take-off an output of 80 MW, equivalent to that of 1 600 000 energy slaves. In other words, to develop all the power delivered by the chemical bonds of the jet fuel, each time a 747-400 aircraft takes off from Malpensa (Milan, Italy), or from any airport for that matter, would require the muscle power of all the inhabitants of Milan and its hinterland.

Finally, a large thermoelectric power plant (800 MW) could operate by "muscle power" thanks to the continued work of over a quarter of all Italians: that is, 16 million people. Electrical utilities in Italy can deliver up to 120000 MW of electrical power, which is equivalent to the human muscle power of 2.4 billion people.

From Coal to Coal?

Other than to meet basic energy needs like cooking food and heating, timber has always been used as a raw material for building houses, ships, and artifacts of all kinds. In 1891, the United States population consisted of 31 million inhabitants; 90% of the energy was obtained from timber that soon began to run low because of deforestation of this abundantly available resource. For centuries, progressive deforestation also took place on the European continent; at one point it became unbearable.

Today, deforestation continues in other regions of the world: it is estimated that each year around 160000 km^2 of forests are being destroyed, an area equal to more than half that of Italy. European countries are among the largest importers of lumber.

Exploitation of coal, a source of energy more difficult to access (it had to be mined) but much more abundant, began in England between the sixteenth and seventeenth century. Coal had been known for some time, but until then it had not been exploited to any great extent because it was unattractive and had to compete with the abundance of timber.

If we compare equal weights, coal is a much more powerful fuel than wood. With coal it was possible to obtain a much greater quantity of useful work. The demand for coal grew continuously to the point that it had to be mined at greater depths. At the beginning of the nineteenth century, mines as deep as 300 meters (1000 feet) were not uncommon. On this point, it's important to remember that the living conditions of miners, often women and children, were for the most part unbearable. The human cost of this mining activity was enormous (and still is in some countries). Centuries ago, mankind began to experience the damage caused by the consumption of fossil fuels – not just benefits.

The ever-increasing availability of coal led to an increase in the availability of processed metals obtained from melting processes in high-temperature furnaces. Thus began the era of machines.

The most important innovation made possible by the abundance of coal was the steam boiler patented by James Watt in 1769 in England. It converted the chemical energy of coal into thermal energy and then into mechanical energy. Consequently, after millennia, human and animal muscle power and the energy from windmills and watermills could easily be replaced by powerful machines. It was the beginning of the industrial revolution!

At the end of the nineteenth century, the use of coal as a fuel had exceeded the use of wood and agricultural wastes to fuel industries. England and the United States were the countries with the highest production of coal, and were therefore at the forefront of the transition from a traditional artisanal economy to one of industrial production.

The rudimentary and inefficient Watt steam engine was gradually perfected. At the end of the nineteenth century, steam boilers were 30 times more powerful and 10 times more efficient than the models available early in the century, although they were still too heavy to be usable for road transport.

In 1900, coal was the source of 95% of commercial energy – then came the era of oil. The first to extract oil were probably the Chinese in very ancient times. Industrial extraction of this *black gold* began in the United States in 1859 at Oil Creek in Pennsylvania; however, the first commercial extraction of crude oil in North America occurred in 1858 in Canada's Old Ontario Oil Belt (see Chapter 11). In the second half of the nineteenth century, oil extraction also developed in other areas of the United States, most notably in Texas and California, as well as in Romania, on the Caspian Sea, and in Indonesia.

In the first decade of the twentieth century, oil drilling began in Mexico, Iran, and Venezuela. The first oil well in Saudi Arabia began production only in 1938. After World War II, we began to see exploitation of deposits of another fuel in consistent fashion – natural gas – whose major property was, first of all, its relatively minor impact on the environment. Natural gas has proven to be a valid replacement of coal and oil in several applications.

The development of means of transport began with oil and with the invention of the internal combustion engine. The watershed between coal and the era of oil was in 1911, when England decided to convert its naval fleet from coal-fired power to oil power.

Unlike the timber era, which ended for lack of raw materials, the era of coal began to decline not because of a lack of coal, but because a more valid alternative became available. In reality, however, the era of coal, especially in industrial processes, is not over yet. Even today coal provides approximately 25% of primary energy, most of which is used to produce electricity. Because of the scarcity of petroleum, this percentage is likely to rise. China, which has abundant coal reserves, extracted about 3500 million tons of coal in 2011. It is expected that production will increase by 4% annually until at least 2030, when it will be more than double current levels.

Hidden Energy

There is a "hidden" energy in all things that surround us. An analysis of the energy cost of a product is a complex operation and often involves questionable parameters. The values reported below are therefore approximate estimates. Nonetheless, they do clarify a fundamental concept: to produce anything useful (but equally, anything harmful) takes a lot of energy. For instance, it is estimated that to produce a tonne (t) of paper sheets requires an amount of energy equal to 0.8 toe (tonnes of oil equivalent); for the production of plastics, 1.5–3.0 toe/t; for aluminum, about 5 toe/t; for titanium, a metal widely used in the aerospace industry, approximately 20 toe/t.

To manufacture a car necessitates, on average, 3 toe/t. Thus, we can estimate that even before the car begins to circulate, it has already consumed approximately 25% of its total energy consumption before its relegation to the scrap yard. To manufacture a computer requires a quantity of energy equivalent to that provided by approximately 250 kg of oil. This means that, even before the computer is turned on, it has consumed a quantity of energy about three times greater than that which it will use throughout its useful life. Scrapping such products means throwing away the energy that was used to produce them in the first place.

Obtaining energy also costs energy. For instance, extracting oil from the rich wells of the Middle East costs 5% of the energy extracted, whereas to produce oil from Canadian tar sands (see Chapter 11) requires energy expenditure of up to 35%. Transportation of the black gold by oil tankers costs around the equivalent of 1% of the energy carried.

Similar transport of natural gas by tankers costs 10–15% of the energy they carry, since the gas must be liquefied first at −162 °C and then kept in that state for several days before reaching a re-gasification plant. From there, it enters the network of pipelines, wherein it travels through special pumping facilities located along the way about every 100 kilometers (about 60 miles) or so. Transport via pipelines has an energy cost three times more for gas than for oil.

Mining for low-quality coal can cost 20% of its energy content. Other energy is required, of course, for work-up processes, for transport, and ultimately its conversion into other forms of energy.

The production of electricity by a conventional coal power plant has a limited efficiency: such production exploits only 30–40% of the primary energy. Other energy must be spent in the construction of the power stations, the transmission systems, not to mention the further losses of energy that occurs in the transmission of electricity over long distances.

In the case of photovoltaic panels, wind turbines, and other renewable energy sources, the cost of transporting the energy source is zero since the energy generated is used locally. However, there is an energy cost in the manufacturing of photovoltaic panels as with any conventional technology. It is estimated that the amount of energy used to build photovoltaic panels can be recovered in 1 to 3 years of operation. As we will see later, this *payback time* is higher than that calculated for wind turbine farms, but much lower than what is typical for nuclear power stations.

From Faraday to Blackouts

On 17 October 1831, the English scientist Michael Faraday demonstrated that it was possible to transform mechanical energy into electrical energy and *vice versa*, using the phenomenon of electromagnetic induction. According to legend, during a visit to his laboratory, the British Prime Minister asked what was the purpose of this *exotic new substance* called electricity, to which Faraday replied – *don't worry, my Lord, one day you will tax it.*

The ability to generate mechanical movement from the combustion of coal, oil, and gas, combined with Faraday's discovery, made possible the large-scale production of electricity, which was also aided by the development of a new type of engine: the steam turbine.

The first power station came into operation in London on 12 January 1882. The development of electricity was due, in good measure, to the genius and to the inventive and entrepreneurial capacity of a handful of engineers and scientists: Thomas Alva Edison, Nikola Tesla, and George Westinghouse are noteworthy.

Other than being of fundamental importance for many industrial processes and for rail transport, electricity has changed the face of cities and the lifestyles of citizens of the richest countries. Electricity enters our homes clean and quiet in a continuous and precise fashion. It allows us to turn on the lights and to operate our refrigerator, washing machine, TV, radio, telephone, air conditioner, computer, and dozens of other devices that make our lives less strenuous and more enjoyable. All this is possible thanks to the existence of the most extensive and complex infrastructure ever built by man: the International Electrical Network, which for Europeans extends from the Atlantic to the Urals and from the Arctic circle to North Africa.

Although we hardly realize it, it is thanks to the available electrical energy that we are able to store food for days, wash our clothes without effort, see what's happening around the world, listen to music from a stereo, write a text without using pen and paper, and do many other things that have become ubiquitous habits in our daily lives.

Our dependence on electricity has reached pathological levels, however. Without it, we can no longer carry out virtually any activity, as experienced by the several blackouts that occurred in the United States, in Italy, and elsewhere in Europe and Asia.

We mustn't forget, however, that 1.4 billion people have no access to electricity. As we shall see later, their only hope of having access to electrical energy is by means of the widespread use of small-scale technologies of renewable energy.

From Muscle Work to Jet Aircraft

Until the nineteenth century, travel and transport of goods were done in large part by muscle work of men and animals, and by the occasional exploitation of wind power at sea and power from river currents.

Horses, oxen, camels, and elephants were the animals used most often. Roman chariots pulled by oxen could carry up to 500 kg of cargo a distance of 15–20 km. Caravans of camels carried salt from the Bilma salt marshes to Agadez (610 km) through the Nigerian desert Ténéré, traveling over 40 kilometers (25 miles) a day with a load of nearly 100 kg per head. These caravans have now been supplanted by trucks, each of which can carry a load equal to that of 250 camels.

The first major change in the transportation sector came with the invention of Watt's steam boiler. At the beginning of the nineteenth century, with improved efficiency and reduced size, coal-fired steam boilers were used for the first time on river boats. The first crossing of the Atlantic Ocean by steamship took place in 1833 along the Quebec–London route.

Likewise, Watt's steam boiler led to the development of rail transportation. In 1825, the first rail of George Stephenson's Stockton & Darlington Railway was laid on Bridge Road in Stockton-on-Tees (UK); it was the world's first passenger railway connecting Stockton to Shildon. Five years later (1830), the first public railway system began to operate in England between Liverpool and Manchester. The first railway line along the Napoli–Portici route in the Italian peninsula was inaugurated October 3, 1839 by the Bourbon Duke Ferdinand II. The year 1904 saw the establishment of the Trans-Siberian Railway, which was and remains the world's longest railway line (about 9000 km) connecting Moscow to the Siberian Far East. Today, most trains run on electrical power. In some countries, trains run at speeds greater than 300 km/h.

Despite the enormous progress, the steam boiler engine did not dominate the field of transportation, as the middle of the nineteenth century witnessed the birth, use, and diffusion of a formidable competitor to coal, a fuel much more flexible, easy to transport, less polluting and more powerful: crude oil (petroleum).

The availability of petroleum gave impetus to the development of internal combustion engines. With this new technology, it was no longer necessary to produce steam to generate mechanical power as was the case for coal. Liquid and gaseous fuels can produce work directly in the engine's combustion chamber where it burns without the intervention of steam as an intermediary. Internal combustion engines are inherently much more efficient than steam boilers.

Patents issued in the field of internal combustion engines were many. The first was issued in 1857 to Barsanti and Matteucci in Italy. Then came those on the four-stroke engine (Otto, 1876), on the vertical cylinder engine (Daimler, 1885), and the diesel engine (Diesel, 1892). In 1885, Karl Benz built Germany's first four-wheeled vehicle, sparking the artisanal production of rudimentary cars.

The first manufacturer of small cars emerged near the end of the century. The Italian company FIAT (Fabbrica Italiana Automobili Torino) was founded in 1899. In those years, public transportation in the cities was still performed by horses; in 1901 there were about 300 000 horses in London alone. The industrial large-scale production of cars began in the United States due to the initiative of a young entrepreneur from Michigan – Henry Ford – who put on the market the first car at affordable prices: the Ford T. In a few years, horse dealers who had greeted the first car with smiles and compassion had to change job.

On the morning of December 17, 1903, on the coast of North Carolina, a machine invented by man took off in the air for the first time and flew for some tens of meters. Two mechanical geniuses of Ohio, the Wright brothers – Orville and Wilbur Wright – had managed to realize a dream that for millennia had fascinated scientists and artists such as Leonardo da Vinci during the Renaissance period.

The development of aviation figures as one of the most important technological advances of the twentieth century, together with the beginning of space exploration. The first three decades of the 1900s saw the emergence of several prototypes of aircraft. Aviation history of those years is full of memorable successes and sensational failures.

The first scheduled commercial flight from Paris to London took place in 1919. Only after the Second World War did civil aviation establish itself fully with the development of jet aircraft and the progressive manufacturing of lighter aircraft construction materials. The first jet, the Boeing 707, entered service in 1958. In a matter of a few years later, planes took over intercontinental travel that was done earlier by steamships. Most modern aircraft, like the Airbus A380, can carry several hundred passengers. It's now possible to reach any place on Earth by air travel with non-stop flights in less than 24 hours.

Insofar as goods and products are concerned, market globalization has significantly increased traffic by sea. The amount of goods transported worldwide by ships today is four times greater than that carried by trucks, and 400 times greater than that carried by air cargo. Energy consumption for transport by sea is half that for road transport, but 20 times greater than by plane. In the latter case, however, the economics are drastically inverted: to transport just one load of a large ship would necessitate several hundred huge cargo planes.

Means of transport have become the icon of the developed world. It is estimated that the total distance traveled annually by passengers in various types of transport is in the quadrillions of miles per year, 85% of which is by car. The negative consequences of this unstoppable expansion of car travel are traffic accidents, air pollution, and climate change.

Petroleum to Food

Before fossil fuels were put into use, most of the work force of a country focused on agriculture. In the United States, the world's largest producer of food, farmers (agricultural labor) now represent less than 1% of the work force. In modern agriculture, the work of man and animals has been supplanted almost entirely by energy supplied by fossil fuels. These fuels are used in the fabrication and operation of agricultural machinery, in land irrigation, in the production and distribution of fertilizers and pesticides, in the preservation of crops, and in their work-up and transportation.

In the course of the twentieth century, the extent of cultivated lands has grown only 30%; however, the amount of crops harvested has increased 6 times. This has been made possible thanks to a 150-fold increase in the energy used in agriculture, primarily fossil fuels and electricity. Today, the world's crops feed about 4 people per hectare, whereas in 1900 they fed only 1.5 people per hectare.

Clearly then, even agriculture has become oil-dependent. The energy cost is 0.1 toe/t for wheat and 0.25 toe/t for rice. For other agricultural products, the relationship between energy contained and energy consumed to produce them is even more unfavorable. For example, greenhouse vegetable products have an energy content of up to 50 times less than the energy used to produce them. Also, it has been calculated that to raise a 5-ton cow requires energy expenditure equal to 6 barrels of oil (about 1000 liters), and to produce 1 kg of beef takes 7 liters of oil.

From Fire to Air Conditioning

Cavemen warmed themselves using fire. For millennia, lumber was the energy source to cook food and to protect oneself during cold periods. Even today, about a billion people have only wood, dung, and dry scrub for cooking and for keeping warm. If we exclude South Africa and the Mediterranean countries, biomass still represents 70% of primary energy consumption in Africa.

Biomass, which consists mostly of wood, plant debris, dried dung, and other natural materials and wastes, is used as a fuel in poor countries. The weaker fraction of the population, primarily women, spends most of the day in the environment in search of energy and water resources for domestic consumption. For many millions of people, the prospect of pushing a button or opening a water faucet to obtain energy or water without fatigue remains a distant dream.

Since the Renaissance, houses of the nobility and wealthy Europeans were heated with wood stoves. With time, these objects have become works of art to be admired in various museums. With the advent of coal, it was possible to build more powerful boilers, and with the help of electric motors came forced air circulation and hot water, and ultimately centralized heating systems.

Because of the devastating environmental impact of smoke (have you heard the expression *London smoke*?), coal was gradually replaced first by oil and then by natural gas. When the cost of electricity began to be more affordable in the middle of the twentieth century, and with the greater wealth of the population, air conditioners (ACs) began to spread in the United States, first window ACs for single rooms, then as centralized AC systems. The use of air conditioners has rapidly expanded in rich countries and in areas of the world with hot and humid climates. Most cars today are equipped with air conditioning.

The widespread use of air conditioning in buildings has profoundly changed the distribution of electricity consumption during certain periods of the year. Until a few years ago, peak electrical consumption was recorded during the colder months and during the darkest days of winter. Now, however, this consumption also occurs in the hottest summer months.

Energy consumption for heating or cooling a house depends on many factors, such as the orientation of the house, the architectural design, wall insulation, and even more so the type of windows. A study conducted in Germany showed that in

1970 a single-family house of 100 square meters (about 1100 square feet) consumed about 3500 liters (or about 1000 US gallons) of fuel per year; After the provisions for thermal insulation of 1982, consumption fell to 1700 liters (485 US gallons), and then to 1000 liters (about 285 US gallons) after the additional provisions of 1995. With modern technologies, energy consumption can drop even further to about 500 liters (140 US gallons). In temperate regions, it is even possible to build *passive houses* that do not consume energy, but use only sunlight for illumination and heating. Currently, there are about 20 000 such passive houses in Germany.

Even though the structures of homes have improved, there is always the risk that the energy consumed domestically will continue to increase. This depends on the explosive growth of electrical energy-consuming appliances and gadgets. While the efficiency of some appliances (e.g., refrigerators and washing machines) and light bulbs is on the increase, the efficiency of others is on the decrease; this is the case for telephones and plasma televisions. The most illogical waste of energy is the electricity consumed by appliances and gadgets placed on *standby*, that is, switched on but not in operation. This will be taken up in the next chapter.

From Horseback Messengers to E-mails

In the last few years, nothing has changed as fast as the quantity and speed of propagation of information – the information highway. A California university professor estimated that a single issue of the New York Times contains more information than a contemporary of Shakespeare could have collected over the course of his lifetime.

In the 1800s, as in Roman times, information was transmitted by travelers or – in very special cases – by messengers on horseback who could at most travel 200 km (about 120 miles) a day. The news of the victory of Admiral Nelson of October 21, 1805, over the Franco-Spanish Armada at Trafalgar, West of Gibraltar, arrived in Britain nearly 2 weeks later on November 4 of that year. In 1865, news of the assassination of President Lincoln of the United States reached Europe only a week later. In 1953, the manuscript describing the work of Watson and Crick on the structure of DNA took more than a month to reach the editors of the journal *Nature*. Today, scientific articles are being submitted to publishers online via Internet in a matter of seconds.

All of this of course has had a direct influence on costs. For example, the cost of sending a telegram overseas in the 1950s was 20 cents per word. Today an e-mail message as long as one wishes can be sent from one end of the world to the other in an instant, paying a relatively small monthly fee to a service provider for connection to the Internet.

The production and distribution of an enormous amount of information in real time is one of the distinctive features of modern society. What today is called *the information society* has matured into a crescendo of inventions that have characterized the last 130 years: telephone, radio, cinema, television, transistors, computers, satellites, telecommunications, internet, and e-mail, among others.

Today we can send real-time information anywhere in the world by simply typing on the keyboard of our personal computer. This apparent very simple action has been made possible by electrical energy. Energy efficiency in this field is also rapidly improving.

The first electronic computer (the ENIAC of 1944) consumed 200 kW. Today, a desk-top computer uses less than a thousandth of that power; a laptop requires only 50 W. However, consumption has increased exponentially because there are now hundreds of millions of computers with printers and copiers (and so on) around the world. In addition, energy is consumed by our computers not only to transmit information, but also to maintain the huge complex international tele-communications network so as to enable fast and reliable exchange of packets of information at any time.

Whether this information is useless or not, wrong, confusing or misleading, is another problem that someone has summarized as: *At one time we sought wisdom, then we were happy with knowledge, now we're left only with information.*

From Gunpowder to the Atomic Bomb

In the last 150 years, the widespread availability of low-cost energy has also served, in an impressive way, to increase the world's destructive power of military arsenals bringing humanity to the brink of self-destruction.

Even though the cold war ended over 20 years ago, the fact remains that both Russia and the United States retain nearly 10 000 nuclear warheads as part of their military arsenals.

Energy plays a triple role in the economy of war. Large amounts of energy are needed to produce weapons (rifles, guns, tanks, planes, ships, missiles) capable of shooting at the enemy other forms of energy *packed* in the most concentrated and most devastating possible manner (bullets, explosives, incendiary bombs, chemical weapons, nuclear). Once the war is over, of course there comes the need to rebuild the devastated countries' infrastructures at even greater energy costs.

Weapons used during the middle of the nineteenth century were not much different from those available in the eighteenth century. New explosives (dynamite, cyclonite), much more powerful than traditional gunpowder, were developed between 1860 and 1900. The contemporary development of metallurgy by the use of coal led to the production of better quality steel. This permitted the range of guns to increase from 2 to 30 km (i.e., from about 1 to 18 miles) in the years between 1860 and 1900.

The first warships and submarines were launched in the early 1900s. On the eve of the First World War, Britain decided to convert its fleet from coal-fired power to oil power. Since then, oil has become the most important strategic material of the military establishment.

Military aircraft made their appearance on the battlefields of World War I, where chemical weapons (mustard gas, chlorine, and phosgene) were used on a large scale for the very first time, especially on the Franco-German front, causing hundreds of thousands of casualties. Development and fabrication of tanks, fighter aircraft, bombers, and aircraft carriers took place within a few years during the period between the two World Wars.

On the 6th and 9th of August 1945, respectively, the two Japanese cities Hiroshima and Nagasaki were razed to the ground with the first nuclear weapons produced – the atomic bomb. The end of the Second World War saw the world divided into two blocs: the East and the West. This led to an unrestrained arms race to develop new technologies for destruction, such as long-range bombers, nuclear-powered submarines and aircraft carriers, and intercontinental missiles: all war machines that carry nuclear warheads with a destructive power thousands of times greater than those used on Hiroshima and Nagasaki. It is estimated that from 1940 onwards, approximately 10% of all energy used in the world has been used for the development and the fabrication of weapons.

The firepower deployed has grown progressively: in 1914 the British Royal Air Force had about 154 airplanes. During World War II, the United States produced more than 250 000 warplanes.

The most massive attack of World War I employed 600 tanks, while in the final assault of World War II the Soviets used 11 000 tanks, 8000 aircraft, and 50 000 cannons and missile launchers against Berlin.

In 1944, the Allies could field firepower three times greater than that of the Axis Powers, who were desperately short of energy resources to feed their war machine. This was mostly responsible for reversing the tide of the conflict.

To defeat the Nazi regime, the Allies unloaded a total of about a million tons of bombs on Germany. The economic costs of the deployment of this power were enormous: in 1944 the United States and the Soviet Union used, respectively, 54% and 75% of their gross domestic product (GDP) for military expenditure.

The human costs of the destructive power of the weapons of war have increased progressively: the most bloody battle of the First World War (Somme, 1916–1918) caused about a million casualties; the battle of Stalingrad – now Volgograd (August 22, 1942 to February 2, 1943) – was among the bloodiest battles in the history of warfare claiming the lives of more than 2 million people, not to mention the nearly 900-day siege of Leningrad (now St. Petersburg) from September 8, 1941 to January 27, 1944 that has been described as the most lethal siege in world history and caused nearly 1.5 million casualties (soldiers and civilians) and the evacuation of 1.4 million people (mostly women and children) many of whom died during evacuation due to starvation and bombardment. The bombing of Germany by the Allies caused 600 000 casualties; the two nuclear bombs dropped on Japan exterminated in a few moments more than 100 000 civilians. The fallen in World War II amounted to about 55 million people, 70% of whom were civilians.

The carnage ended with the unconditional surrender of Germany and Japan in 1945. Since then, there have been continuous trickles of small and large conflicts that have caused millions of casualties. To give you an idea of the resources employed today in the development of weapons and the waging of war – euphemistically called "defense spending" – in the year 2011, the world's total military spending was estimated to be $1738 billion, representing 2.5% of global gross domestic product or $249 for each person.

Emerging Issues

With this roundup we have attempted to give you – the reader – an idea of how all aspects of our lives depend on energy. It needs to be emphasized that the major changes mentioned have not affected all nations of the world, and that not even all citizens of so-called *advanced nations* have benefited from this energy abundance. A spiral of *availability of energy – technological development – wealth – energy consumption* has been created that has led to great inequalities between people and nations. Such inequalities will be hard to overcome. For example, with approximately 330 million inhabitants, the United States has about 842 motor vehicles per 1000 people, babies included, whereas in China and India, with a total population of about 2.5 billion there are, respectively, 40 and 20 vehicles for every 1000 inhabitants.

To remedy such inequalities would require putting at the disposal of developing countries an enormous amount of energy. For instance, if China and India had 842 vehicles per thousand inhabitants, assuming an approximate average travel of 10 000 km per year (about 6000 miles a year) and a consumption of 7 liters per 100 km (or about 1 US gallon per 30 miles), these two countries would consume about 8 billion barrels of oil a year. This represents 22 million barrels a day, more than twice the production of Saudi Arabia, which produces a quarter of the world's oil supply.

Who will provide the fuel to China? This question, which for years has hovered over the economic and political establishments, has turned into a nightmare, which has led oil prices to increase significantly since 2003. These cost increases, however, have also been caused by the continual insatiable thirst for energy by the developed countries.

The history of the last 150 years teaches us that the increasingly widespread wealth and material well-being in the more advanced countries create new *needs*. Take tourism, for example, almost non-existent until 50 years ago, and now one of the most energy-consuming activities in the world.

The use of fossil fuels to produce energy is very convenient and very useful. It is a real treasure found in the depths of spaceship Earth, a treasure that man has discovered and used extensively. However, in the last 20–30 years this treasure has posed several serious problems. It is destined to run out. Its use causes severe damage to human health and to the environment. Its irregular sites in various areas of the planet have created economic inequalities, political tensions, and even wars.

How can these problems be solved? How can we bridge (at the same time) these inequalities that threaten peace, meet the needs of those who are accustomed to luxury and waste, address the limited availability of fossil fuels, and avoid damage to the Biosphere from the use of fossil fuels?

Big challenges await us. We have to deal with them as soon as possible, before the occurrence of physical events that may become unstoppable and ungovernable, and that could be accompanied by social and political unrest and bring humanity to a painful future. As we shall see later, the challenges are not insurmountable. In fact, they may even turn out to be great opportunities.

3
How Much Energy Goes to Waste?

Good sense was there, but was hidden for fear of common sense.

Alessandro Manzoni

As we saw earlier, the Second Principle of Thermodynamics poses insuperable physical limits to the useful conversion of energy. When energy is produced (i.e., transformed), a fraction of the final product is in the form of heat, which can never be fully used. In a sense then, wasting energy is one of the laws of physics and chemistry.

Unfortunately, the huge availability of cheap energy, which man has enjoyed over the last 50 years, has quietly amplified the amount of energy wasted far beyond the physical constraints, making it one of the main features of our life-style – often a genuine insult to good sense.

For instance, only 44% of the primary energy is transformed into useful energy in the United States – the remaining 56% is lost. The industrial establishment could, in principle, recycle part of this wasted and lost energy to produce a quantity of electricity equal to that of 65 power stations of 1000 MW each. An action of this kind would overcome and silence the troubling discussions on whether the United States should revive its civilian nuclear program that has remained idle for some 30 years.

The Largest Explosion of All Time

The 9th of September is not a commemorative day for any country of the world. And yet, unknown to most people, on the morning of 9 September, 1913, an extraordinary event occurred that has literally changed the course of history. In the laboratories of the German chemical giant BASF at Ludwigshafen in the Rhineland began the industrial production of ammonia (NH_3), a molecule consisting of one atom of nitrogen and three atoms of hydrogen. The chemist Fritz Haber and the engineer Carl Bosch were rightly proud of the success of their studies, and probably never imagined that they had triggered the biggest explosion in history: that of the human population.

Powering Planet Earth: Energy Solutions for the Future, First Edition. Nicola Armaroli, Vincenzo Balzani, and Nick Serpone.
© 2013 Wiley-VCH Verlag GmbH & Co. KGaA. Published 2013 by Wiley-VCH Verlag GmbH & Co. KGaA.

Nitrogen is a chemical element necessary to sustain life, including human life. It is an essential component of the amino acids that the body uses to synthesize proteins. The molecule of nitrogen, N_2, is inert and therefore difficult to use. Not by chance, it is very abundant in the atmosphere and makes up 80% of the air we breathe. Until that day of September 9, 1913, only nature – through various types of bacteria – was able to *fix* atmospheric nitrogen by extracting it from the air to be used in the synthesis of biologically relevant organic compounds.

In the past, man had limited himself in providing a modest amount of extra nitrogen to the soil by recycling livestock manure and other natural wastes, a good source of proteins. Agriculture turned a page on that day: the use of fertilizers produced from synthetic ammonia started the *green revolution*, which led to greatly increased productivity of the soil and availability of food.

The sudden and simultaneous availability of fertilizers and synthetic fossil fuels largely explains the impressive surge in population that occurred in the first few decades of the twentieth century. To make up the first 2 billion people required 5000 years. The next 2 billion people appeared in about 50 years (1927–1974). The last 2 billion were born in just under 25 years from 1974 to 1999. At the last count (November 2011), the world's population was about 7 billion people. It is expected to stabilize around 9–10 billion around 2050.

The daily growth of the population on planet Earth amounts to more than 200 000 units or about 80 million each year – that is, a medium-sized European city every morning and a country almost the size of Germany at the stroke of each New Year. All these people have a right to a dignified existence, with their share of food and energy.

At the dawn of the twenty-first century, the availability of food is not the most immediate problem that mankind faces. What is immediate, however, is its distribution and the ability to afford it economically. Indeed, we have now reached the point of greatly abusing this extraordinary conquest in both rich countries and developing countries. That is, overeating is causing increased harm to human health and to the environment. We must remain vigilant, however, for while man is getting fat there are signs of a gradual impoverishment of the fertile soils of our spaceship, often a result of ill-considered management and amplified by the effects of global warming.

Obese and Miserable

It is hard to imagine, but it has happened. There are now 1.5 billion overweight or obese people, and **only** 1 billion that have problems of food supply. In layman's terms, the obese exceed the hungry.

The increase in obesity is particularly acute in developing countries. In some cases, it reaches almost pathological levels – as in Mexico and Egypt – countries which until recently often had to deal with hunger.

The reasons for this rapid transition from a poor to an over-abundant diet are tied to the increased consumption of meat and vegetable oils, to the spread of sweet drinks instead of water, and to urbanization that has induced sedentary lifestyles; in short, to the prevalence of the most harmful Western eating patterns caused by globalization.

In fact, without going too far, it is almost unreal that in a country like Italy, which until a few decades ago grappled with rickets and pellagra, today has gyms where people sweat to get rid of their fat (overweight) and clinics that practice liposuction. In less than two generations, our relationship with food has passed from a search regime to one of defense. With some good justification, there are those who study survival strategies to offset the binges of Christmas festivities.

Every calorie we eat today requires on average one calorie of fossil fuels to bring that food to the table. There are those who claim, with some reason, that modern agriculture is nothing more than an industry that converts fossil fuels into food, and unfortunately too often it does so inefficiently.

Fruits Out of Season

The availability of cheap fossil fuels has completely distanced the food producer from the consumer. For a long time and until only a few decades ago, the producer and the consumer often coincided (people produced their own food) or had a direct relationship. However, no one seems to show any nostalgia for those times, often of hunger and deprivation. Nonetheless, we need to acknowledge that the abundance of energy has created a system of food distribution that is now in some aspects somewhat perverse.

For example, we should not be surprised to find apples from China, green beans from Egypt, oranges from Chile, and kiwis from New Zealand at the supermarket—all products that could easily be grown locally in a country like Italy. Instead, we get them from other countries, indeed from other continents, consuming large quantities of fuel and discharging millions of tonnes of carbon dioxide and pollutants into the environment. This is the hidden cost of our senseless greed for out-of-season products. Apart from their questionable flavor, we should reflect on the loss of important values, especially for the children: the perception of seasons, nature's cycles, and, more generally, the sense of limits.

When someone seeks your advice on how to save energy, suggest he refrain from buying, for example, imported strawberries at Christmas. At first he will be rather puzzled; then, on reflection, he will come to understand that it is one of the best things he can do to change an energy system that we can't afford to continue to use forever.

From Whale Oil to Pollution by Light

For millennia the rhythm of people's daily lives was marked by a series of daily cycles: light–darkness–light. At sunset, nearly all human activities were interrupted because of the inability to continue in the dark. Until the end of the eighteenth century, lighting a city at night was largely entrusted to torches placed in front of shops and taverns, and to the votive lamps of holy images. The situation was not very different from that of the Middle Ages, when towns took steps to prevent possible and hidden dangers—at sunset people retreated indoors, town gates were shut, and curfews were set in place during the night hours.

We must also remember that houses did not benefit from insulating and perfectly transparent glass, as we have today. Archeological evidence from Roman patrician houses have shown, for example, that during the day in cold months even the Eternal City (Rome) was forced to choose between incoming light and outgoing heat through the windows. It was not possible to have a warm house and at the same time a house flooded with light.

For millennia, man has used fire to keep warm, to cook food, and to illuminate the night hours. For the latter, he gradually learned to tame the flames by the use of torches and lanterns. Historical and archeological examples of the latter of great artistic merit exist from all civilizations.

Starting with coal in the seventeenth century, massive exploitation of fossil fuels began to revolutionize not only the productive system but also the daily life of people. In this regard, the use of *city gas for public lighting* in the early decades of the nineteenth century constitutes an authentic icon of the industrial and technological revolution.

Up to the middle of the nineteenth century, tens of millions of rural homes worldwide relied on fuels of animal and vegetable origin for artificial lighting, sometimes including products as exotic as whale oil and beeswax.

Only later, in 1879, Thomas Edison patented the carbon filament lamp. This device was 20 times more efficient than a candle, which converts a measly 0.01% of the chemical energy of the burned wax into light. Edison's device, developed during the years in which electrical technologies were being developed, was a crucial step toward a modern lighting system. However, it was still too inefficient (0.2%) at converting very expensive electricity into light.

Incandescent tungsten lamps introduced and perfected in the first decade of the twentieth century (1906, still in operation today) were followed by the yellow-orange sodium vapor lamps in the 1930s (very efficient, now the standard for public lighting), by fluorescent tubes in the 1940s (often mistakenly called *neon lights*), and by halogen lamps in the 1960s. Those that are currently described as *energy-saving lamps*, technically *compact fluorescent lamps* (CFL), were introduced around 1980—they have an efficiency of light conversion 50 times greater than Edison's light bulb.

Thanks to artificial lighting systems today, the day fades into the night without us even noticing it. All human activities continue regularly. Freedom from darkness is without doubt one of the greatest achievements of human civilization. However, in this case also, as was the case for food, we have been fooled. In fact, the use of artificial light has become so pervasive that we have reached the point of talking about *light pollution*, a phenomenon that limits viewing the stars in the most densely populated areas of the planet. This pollution also has adverse effects on plants and migratory birds, altering their vital rhythms.

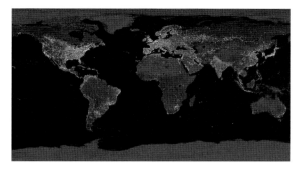

Figure 2   The Earth at night seen from space – nice, but so much light wasted! Image courtesy of NASA (http://www.visibleearth.nasa.gov).

This modern form of pollution is easily illustrated by the images of the Earth seen at night from space (Figure 2). This suggestive *collage* of satellite images shows that the richer and more populous regions of the planet shine during the night, whereas the poorest and most remote and uninhabited regions are immersed in total darkness. Comparison with similar pictures of 10 or 20 years ago shows a strong expansion in artificial lighting.

Geographically, this trend overlaps with maps of economic growth. China, India, Southeast Asia and Eastern Europe are today much brighter than they were in 1990. Africa is still almost completely dark. In fact, the evolution of the image of Earth at night shows the trend of economic growth and, in the final analysis, increased energy consumption. However, this information also tells of a giant waste, for the sky does not need to be illuminated by man. The light we see from space is essentially light (energy) wasted.

It would be inspirational if, in years to come, pictures from space could tell a story of human development in an equally clear manner, but with images that change with time in the opposite direction. That is, an Earth which is progressively turned off at night because we have learned in a more efficient and rational way to overcome the darkness. Amateur astronomers would be thrilled. So would migratory birds.

At Full Throttle

We have learned other things by photographing the Earth from space. First, we have come to realize that the blue planet is surrounded by an endless immensity of dark, empty, and inanimate matter. If man were not stupid, he would do well to keep this wonderful *prison* in good health. Among the much useful information that we have obtained from space exploration, there is at least one salutary discovery that illustrates our insane propensity to waste colossal amounts of energy.

Artificial lighting that we observe in satellite photos is yellowish and covers large areas of the Earth. Detailed observations, however, tell us of the occurrence of small, very intense red patches in some particular areas of the planet – the Persian Gulf, Siberia, Venezuela, and the Gulf of Guinea. They are the flames we often see from major oil fields where gigantic quantities of gaseous fossil fuels are burned, as such gases tend to hinder oil extraction.

In the last two decades, we have come to burn in this manner between 150 and 170 billion cubic meters of gas annually. It is an immense quantity, equivalent to 30% of the consumption of the European Union countries, 25% of that of the United States, and 75% of the exports of the world's largest gas producer, Russia. These numbers bear witness, in uncompromising fashion, to the aberrations into which we have fallen in the era of easy and cheap oil.

A Desperate Case – the Transportation System

Traveling on the highways of the United States for Europeans is a unique experience: wide lanes, low traffic density (except around large cities), strict and highly respected speed limits, infinitely straight roads, and very few tunnels. Riding *coast-to-coast* on an immense gasoline-powered four-wheel drive (4WD) sports utility vehicle (SUV) with a 6000 cm^3 engine is a recurring holiday dream in the minds of many Europeans. The problem is that these motorized cowboy vehicles in North America are also found in the narrow medieval streets of Italian and other European cities. These are not places for chasing the American dream, but practitioners are usually quite proud and happy to do it. These examples are among the most glaring examples of a transportation system that has reached its terminal stage, the one that lurches between paralysis and total collapse.

Vehicles circulating in the United States consume about 5% of *all* the world's *primary energy*. It is one of the most unsustainable extravagances of a civilization based on cheap oil. This abnormal consumption can only be partly explained by the fact that average distances in the USA are much greater than those in Europe. Unfortunately, all attempts to pass legislation requiring an increase in efficiency of motor vehicles in North America, which since 1985 has remained at an average level of 11.5 km/liter (or about 25 miles per US gallon), has found only staunch opposition. Starting in 2011, more stringent limits have been set by recent United States Administrations such that in 2016 the average car efficiency should reach 16.6 km/liter (36 miles per US gallon).

This transition to cars with an average consumption comparable to that of cars circulating in Europe and Japan should significantly decrease energy consumption without undue sacrifices for the American motorists. In the meantime, domestic production of oil in the United States has been until recently in continuous decline since the mid-1980s (in the last couple of years this trend has been reversed), while imports continue to cover nearly two-thirds of its needs, which increase concomitantly with military expenditures related to the control of the energy corridor. Anxiety to cover the ever-increasing demand for crude oil has crept into some oil

Car	SUV	train	airplane

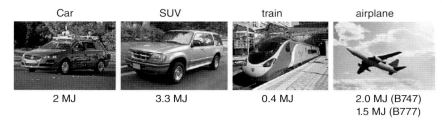

2 MJ	3.3 MJ	0.4 MJ	2.0 MJ (B747)
			1.5 MJ (B777)

Figure 3 A comparison between the performances of certain means of transport, expressed in terms of energy used per kilometer per passenger. Adapted from V. Smil, Energy: *A Beginner's Guide*, One World, 2006.

company CEOs, who through expensive advertising campaigns have promoted the imminent end of cheap oil.[1]

The average European citizen uses nearly a third of his total energy consumption to travel by some form of motorized transportation. Of course, many Europeans love to hike and bike: indeed, moving around on foot or by bike in many cities has become faster and cheaper than doing it by car.

Certainly the pedestrian option is very efficient from the point of view of energy: to move a kilogram of flab with legs costs around 3.5 kJ of energy per kilometer, and to move it by a car of medium size takes about 30 kJ/km. This is not surprising as there are tons of metal, plastic, glass, and fuel with us in the car: among others, there is a 150-horsepower engine, 4 robust security bars, 2–3 square meters of striking chrome, a battery, a radio and six *rave party* speakers, and even a spare wheel which we carry around hoping never to have a need to use it. Sometimes we don't even know where the spare wheel is hidden.

Energy consumption per passenger for a medium-sized car is about 2 MJ/km; a 4WD SUV consumes at least 60% more (Figure 3). It may surprise you to know that a modern and highly efficient aircraft, such as the Boeing 777, consumes less energy than a car on a per passenger and per kilometer basis. In one day, however, a plane can cover distances equal to those that a car covers in a whole year. Climate and environmental impacts of air transport are increasingly under attack, linked to the long distances and the fact that the pollutants are released at high altitudes, particularly in vulnerable areas of the troposphere.

However, the most interesting data on energy consumption in transportation is the extraordinary efficiency of trains: to move by rail means an 80% reduction in energy consumption compared to the car. Rail systems can achieve efficiency levels (and not just energy) totally unachievable by other means of transportation. For example, the Japanese high speed train, the Shinkansen, carries over 150 million passengers a year at speeds exceeding 400 km/h, (ca. 240 miles per hour) and with an average delay per train of 5 seconds or less (you read that right: 5 *seconds!*). The efficiency of this system can also be evaluated on the basis of safety.

---

1) See for example Chevron's website: www.willyoujoinus.com. This site also presents a measure in real time of the distressing world oil consumption.

The Shinkansen has been in operation for over 20 years in Japan and has transported several billion passengers without a single fatal accident. It is instructive to compare these data in parallel with others – road accidents in the European Union currently cause each year more than 30 000 casualties and over a million injured, many of whom are handicapped permanently. The victims of this war are regularly sacrificed to the Gospel of never-ending economic growth. No category, corporation, union, or political party is exempted from preaching this Gospel.

The tens of thousands of deaths on the roads make for thriving business: energy companies, car manufacturers, car mechanics, nursing homes, funeral homes. The gross domestic product increases. This is enough to continue to perpetuate an inefficient and sick system.

If someone suggested quietly, with data in hand, that the transportation system is in a desperate condition, he would immediately be branded a brainless moron who opposes the unstoppable progress of human civilization. It will not take centuries or decades for the wind to change. In Beijing (China) and Bangalore (India) car sales continue to increase exponentially. In the meantime, the CEOs of major oil companies say, candidly, that if we continue on this road they have no idea where the extra 15 million barrels of oil that would be needed daily will come from between now and the next 10 years.

If we don't change the transportation system, it will be the transportation system that will change us.

Let's Get a Move on

We have described some examples of the many ways that energy is wasted and that characterize modern civilization and the normal lives of people. At this point you may ask how you can personally help to reverse this insane course. Tips to save energy and reduce waste can be found everywhere now. Among the most active in these information campaigns are the same oil companies that, worried by the not too encouraging energy projections on the supply of primary energy for the next twenty years, have adopted the strategy of "prevention is better than cure."[2]

The first objective to pursue in the fight against waste is as always *knowledge* – to realize *where* and *how* to consume so as to be more circumspect in our actions. Unfortunately, there is a large disproportion between the perception we have of our energy consumptions and their effective deployment.

As reported in Figure 4, the average European citizen believes that 40% of energy consumption goes to power appliances and lighting systems. In actual fact, these consumptions are 5 times lower, but are much more felt as they are directly *visible*. The reality is that, on average, more than half of the energy consumed by the European citizen serves to heat the environment where he lives, and nearly

---

2) See for example the campaign "Consume better, we all gain" by Italy's ENI company on its website: http://www.30percento.it.

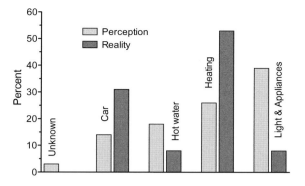

Figure 4 Percentage of domestic energy consumption in Europe. Distribution perceived by citizens and the real consumption. Data source: Euro-barometer 2007.

another third serves to run his car. The first thing to do then is to put on an extra sweater and lower the thermostat at home from 21 to 19 degrees Celsius. This would reduce by up to 20% the energy consumption for heating and would also save the consumer a lot of money on gas and/or electrical bills.

The data of Figure 4 also tell us that we must focus attention on the car. In addition to buying a very efficient one, it would be wise to avoid using it for short routes, reduce highway speeds, and check tire pressures. If we are particularly conscientious with regard to energy saving, we will limit our highway speed to 110 km/h and thereby save 35% fuel.

It is clear, however, that if we use the 100 Euros so saved to buy two *low-cost* tickets to spend a weekend in Tenerife, our virtual energy saved on highways will have been totally ineffective, and, indeed damaging. The final consumption in the use of energy comprises only a 25% share of electricity. However, the European consumption of electricity grows faster than the total energy consumption; thus, even if the energy consumed were lower than perceived, consumption must be kept under control so as to avoid wasting energy.

In many homes, water is heated electrically for shower and bath. This is a waste of resources that is truly absurd from a thermodynamic point of view. In fact, that precious and concentrated electricity comes from a power station that generated electricity by heat produced on burning non-renewable and polluting fossil fuels. Two-thirds of this heat is discarded at the power plant with the remaining third converted to electrical energy.

Another widespread waste of energy occurs when placing appliances unnecessarily on standby, such as televisions, DVD players, audio equipment, and computers (though standby is essential, of course, for antitheft systems and automatic gates). Europe's consumption of this *silent* entry represents 6% of total demand for electricity, equal to half the entire consumption for lighting. When on standby, each of these appliances now requires 5–10 W. However, new technologies should soon reduce this power consumption to less than 0.1 W per device.

With regard to lighting, its inherent efficiency is still low: taking account of the efficiency of electricity production, transmission, and conversion, 100 units of primary energy that entered into a power station produced less than 1 unit of useful energy (light) in a light bulb filament. More than 99 units are converted into heat and spilled into the environment as a residue. This is yet another disturbing waste that is being reduced thanks to the progressive substitution of incandescent lamps with fluorescent lamps, and to the emergence of new technologies for solid-state lighting (LEDs and OLEDs).

Perhaps we'll be able to convince people that it is absolutely necessary to increase efficiencies in the way we heat, light, and travel. However, let's not deceive ourselves. All this will not suffice to address, in any meaningful way, the energy transition that awaits us.

In general, the history of energy and technological development teaches us that improvements in the efficiency of energy conversion are always accompanied by *increases* in consumption, because it also increases the material wealth of people and the ability to purchase newer products. For example, from 2003 to 2007 the electrical consumption in the European Union increased by 6.5%, in line with the trend of Gross Domestic Product (GDP), despite the introduction of standards and technologies that have improved energy efficiency. Accordingly, a simple but unattractive concept must enter the common mindset: in view of the upcoming energy transition, the richest citizens of the planet—ourselves included—must *reduce* their energy consumption, and not just *improve* their consumption.

It is evident that to take our children to school by car is more often than not a wasteful use of energy. However, there are responsibilities that go beyond the sense of citizenship of people that call into question the ruling classes, who have shown themselves in some instances to be incompetent and incapable of making long-term plans. In this regard, how many political candidates give priority in their electoral programs to creating bike paths? They would be a great investment for the quality of life and for the public purse (but not for the GDP): less pollution, less obesity, fewer broken roads, fewer hospital admissions, and lower costs for the national health system.

In the face of this inaction, it is depressing to observe how much zeal and energy politicians and elected officials spend in raising public funds for new roads and Interstate highways to be ready in 15 years, when we'll no longer be able to afford the current energy-intensive transportation system. Without delay, we need to plan for the infrastructures of mass and public transportation, particularly rail. The few available resources must be concentrated there.

In Berlin, a city among the most modern in the world, over the past fifteen years hundreds of kilometers of bicycle paths have been created, following which, traffic decreased by 20%. In California, the richest State of the world's richest country, energy consumption *per capita* is lower today than it was in 1975. This proves that if they wished, the rich could easily consume less energy.

4
Energy in the Spaceship's Hold

The energy from fossil fuels is really the only one that can satisfy the needs of our modern way of life and of our civilization?

Giacomo Luigi Ciamician, 1912

Energy is most useful to us when it is concentrated, transportable, and storable. Fossil fuels (coal, oil, natural gas) meet all three requirements, albeit each in a different way.

Currently, as Figure 5 shows, the seven main sources of primary energy–not to be confused with the final energy–are oil (32.8%), coal (27.2%), natural gas (20.9%), biomass (10%), nuclear energy (5.8%), hydro (2.2%) and renewable sources (0.7%).

Therefore, more than 80% of the energy that the world uses comes from fossil fuels, a limited and a non-renewable resource. An untouched treasure for millions of years in the hold of spaceship Earth, we started to consume it extensively only in the last 100 years. Fossil fuels are a valuable resource, but a non-renewable one. In other words, it can be used only once.

Crude Oil

The use of oil has expanded incessantly since the early years of the twentieth century. In the current historical phase, oil has become the most important energy source and for certain applications such as, for example, producing fuel for aircrafts, it is virtually irreplaceable.

So far, extraction of conventional oil has managed to cope with demand. But how long will this last? A modern proverb from Saudi Arabia says: *My father rode a camel, I drive a car, my son pilots a jet aircraft, his son will ride a camel.*

Translated in scientific terms, this means that a day will come when oil production will reach a peak and then relentlessly diminish, as illustrated in Figure 6. In an economic system that requires ever-increasing amounts of energy, the consequences are easily predictable. If we don't find alternative energy sources in time, we'll see fuel prices go through the roof, as always occurs when goods become scarce; such increases are being felt as we speak: in Italy the price of gasoline is

Powering Planet Earth: Energy Solutions for the Future, First Edition. Nicola Armaroli, Vincenzo Balzani, and Nick Serpone.
© 2013 Wiley-VCH Verlag GmbH & Co. KGaA. Published 2013 by Wiley-VCH Verlag GmbH & Co. KGaA.

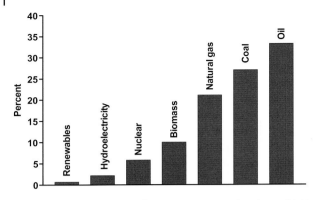

Figure 5 Principal sources of primary energy used in the world. Total in 2009 = 12 150 Mtoe. Source of data: International Energy Administration 2011.

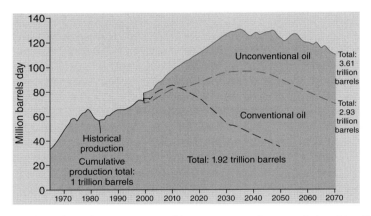

Figure 6 The evolution over time of the world's oil production. According to pessimists, peak production has already occurred in 2010 (lower dashed curve); according to optimists this will occur between 2035 and 2040 (upper two curves). Adapted from Alexandra Witze, *Nature*, 3 January 2007, p. 14.

now around 1.90 to 2.0 Euros per liter (August 2012; in the USA almost 4 US dollars a gallon; in some places even more than 5 US dollars). This will lead not only to economic and political crises, but also to a race to grab whatever energy reserves are available with whatever means possible.

As an exercise or for pure amusement (so to speak), try to imagine what our lives would be like if the flow of oil and gas ever stopped, the flow that quietly feeds incessantly our daily activities. We would all suffer – from the procurement of food to the heating of houses – and more importantly in such details as personal hygiene and the means to get to work (and be remunerated). And what about the perks we now enjoy in our modern lifestyles that we're so proud of? For example, it is not possible to even breathe inside a skyscraper (authentic icon of progress) without the continuous flow of energy (air conditioning, heating, lighting, etc.), since windows are kept sealed. Among the few that still own a piece of land, who will cultivate it without using a tractor?

In this hypothetical situation, which we hope will remain a purely intellectual exercise, we have much to reflect on – the many processes of urbanization of the last 30 years, which were conceived under the wrongly perceived illusion that cheap oil was going to last forever, suburbs that grew very quickly dominated by cars, business parks, and immense car parking lots that have emptied the soul of urban centers and have increased traffic, while nothing has been invested in modernizing and strengthening public transportation infrastructures.

Peaking of Oil Production?

It is difficult to determine when the peak of oil production will be reached. There are certainly sources that have yet to be discovered, but no more large deposits of good quality oil have been found in the last 30 years. Certainly, the supergiant wells in the Cantarell field in Mexico (previously it provided 60% of Mexican production), in Prudhoe Bay Alaska (a major oil field of North America), and the one in Statfjord, Norway (a major oil field in the North Sea) are in continuous decline. Strong doubts have been expressed on the two largest deposits of the Arabian Peninsula: the Ghawar field (about 50% of Saudi production) and the Burgan field (about 65% of the Kuwaiti production). Local authorities maintain an increasingly mysterious reticence concerning their productive vitality. Another certainty is that the search for new oil fields will require large investments. Exploration carries high risks with an eventual return only in the long term, something that international investors have difficulty accepting.

Optimists believe that peak production will be reached in about 30 years, while pessimists believe that the peak was reached between 2005 and 2010. In this regard, a 2010 analysis of oil production of 47 major producing countries by Nashawi, Malallah, and Al-Bisharah (*Energy Fuels*, **2010**, *24*, 1788–1800) of the Department of Petroleum Engineering of Kuwait University estimated that the world production of crude oil will peak around 2014 and that the OPEC production is expected to peak around 2026. The study further noted that based on 2005 world crude oil production and current recovery techniques, the world oil reserves are being depleted at an annual rate of 2.1%.

Of course, there are those who claim that oil is still very abundant and that it suffices only to dig more wells at increasing depths, squeeze the bituminous tar sands and shale (unconventional oil reserves concentrated mainly in Canada and the United States) or get it from coal. These folks fail to mention, however, whether it will be convenient to do so economically and energetically, and whether it will be sustainable from an environmental viewpoint.

If we consider the relentless rise in prices, the two Gulf wars, political tensions between Iran and the United States, and the political instability of the Middle East, it seems difficult to prove the pessimists wrong. And even if the optimists were

Table 4 Energy density in combustible fossil fuels.

Combustible fossil fuel	MJ/kg	MJ/m^3
Coal		
Anthracite	31–33	
Bituminous	20–29	
Lignins	8–20	
Peat moss	6–8	
Crude oil	42–44	30 000–40 000
Natural gas (ambient temperature and pressure)		29–39

Data source: V. Smil, *Energy in World History*, Westview, 1994.
These few data demonstrate why oil is the fossil fuel *par excellence*: a liquid, easy to transport and with a high energy density. Coal and gas are often only awkward cousins.

right, the problem will be felt fully by our children and our grandchildren. Perhaps we may even feel the impact in our lifetime.

Natural Gas

Alternative fossil fuels to oil are its "cousins": natural gas (consisting mainly of methane) and coal. Economically, however, they are but stopgap solutions. In addition to oil being more easily transportable, oil extraction is less costly in energy. In addition, oil has an energy density much greater than that of gas (see Table 4). At ambient temperature and pressure, the energy content of one cubic meter of oil (34 GJ, equal to about 10 000 kWh) is a thousand times greater than that of a cubic meter of gas. This disadvantage tends to limit large-scale use of gaseous fuels in the transportation sector.

Current estimates of reserves suggest that peak production for gas should occur shortly after that of oil. This point also carries much uncertainty, however.

The last ten years have witnessed a sharp increase in gas consumption, caused mainly by the fact that production costs and construction times of gas-fired power stations that generate electricity are highly competitive when compared to other technologies. This has created tensions in the market place because natural gas is also an essential raw material for the chemical industry, which uses it to produce large quantities of widely used materials and substances such as, for example, fertilizers, plastics, medicines, dyes, and pesticides, among others.

Gas production in the United States has increased considerably in recent years thanks to the exploitation of non-conventional deposits of *shale gas*, which consists mostly of methane, and is locked up in schistic rocks usually found more than 1 kilometer deep. It is extracted mostly by the technique known as *fracking*–that is, shattering the rocks with high-pressure injection of water and chemical additives.

Nonetheless, gas consumption is also set to increase, especially for electricity production, owing to the nuclear crisis that was further aggravated by the recent Fukushima incident (Japan).

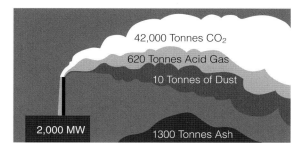

Figure 7 Output of a coal-fired generating station in a single day.

Coal and CO_2 Rise

In view of a possible decline in the availability of oil and gas in the next few decades, attention is being given to coal, whose reserves, until a few years ago, were thought to be sufficient to provide energy for some hundred years. In actual fact, however, coal that is recoverable at low economic and energy cost is not so abundant. According to some estimates, peak coal production will likely be reached around 2050. In any case, even when coal production reaches its maximum, it will provide a quantity of energy less than what we get today from crude oil.

Coal has lower energy content than crude oil and cannot be carried in pipelines, except for short distances. As discussed later, coal is also the most polluting fossil fuel and is the one that generates the most greenhouse gas per unit of energy produced (Figure 7). In this regard, as early as the nineteenth century, the environmental impact of burning fossil fuels was becoming evident. Factories burned coal to power steam engines, steel mills burned coal to make steel, and people in the cities used coal to heat their homes. The poor air quality in many large European cities was far beyond what we can even imagine today.

Estimates indicate that, in 2011, 81% of human-produced energy came from burning fossil fuels, which produces annually around 34 billion tonnes of carbon dioxide (CO_2), a greenhouse gas. It has also been estimated that natural processes of the Earth can only absorb about half this amount, resulting in a net annual increase of about 17 billion tonnes of atmospheric carbon dioxide. The present concentration of CO_2 in the atmosphere (394 ppm) is the highest in the last 800,000 years and the large majority of climate scientists consider it the main driver to the ongoing global warming of the Earth's surface.

The Most Traded Commodities

Deposits of fossil fuels are not distributed evenly in the various regions of the Earth; the same goes for consumption. Table 5 shows the top ten energy consumers; the Italian situation is included for comparison.

Table 6a–c list the ten major countries and the 10 largest producers/consumers of crude oil, natural gas, and coal, with the addition of data for Italy.

Table 5 Total annual consumption of primary energy in the world: the top 10 countries plus Italy.

Country	Total energy (millions of toe)	% of total	Per person (toe)
China	2 613.2	21.3	1.95
United States	2 269.3	18.5	7.28
Russia	685.6	5.6	4.81
India	559.1	4.6	0.47
Japan	477.6	3.9	3.75
Canada	330.3	2.7	9.71
Germany	306.4	2.5	3.76
Brazil	266.9	2.2	1.35
South Korea	263.0	2.1	5.39
France	242.9	2.0	3.16
Italy	168.5	1.4	2.76
WORLD	12 274.6	**100.0**	**1.75**

Source: *BP Statistical Review*, 2012.
The United States, with less than 5% of the world's population, consumes almost a fifth of the world's primary energy (as does China, but which also has a fifth of the world's population). In China, Brazil and even more so in India, energy consumption per capita is still far below Western consumption. In Italy, consumption is moderate, despite widespread wastage. The value of consumption per person is highest in Canada, which consumes the most energy in North America because of its particularly long and severe cold climate.

In addition to the well-known fact that a substantial quantity of crude oil is extracted in the Middle East, it's interesting to note that even though the United States is the third oil-producing country in the world (Table 6a), the first in natural gas production and the second in coal production (after China), it is forced to import about 60% of its crude oil and 6% of its natural gas requirements because of its enormous energy consumption. By contrast, Russia is able to export huge quantities of oil, natural gas, and coal. Interestingly and perhaps not surprisingly, Iran, a nation in the center of complicated international political events in the last 30 years, is the fourth producer of oil and natural gas.

Table 6a–c shows that the most populated countries – China and India – produce far less than they consume. Japan never appears in the top ten producer countries, but is always among the first ten consumer countries. Rich Western countries, except Canada, consume more than they produce. Italy has no significant reserves of fossil fuels, and about 90% of its energy needs come from fossil fuels. It imports 94% of the oil and 90% of the gas it consumes.

Those countries that have no fossil fuels, or not enough for residential and industrial demand, must buy it from producer countries. Oil is perhaps the most traded commodity in the world trade. It is estimated that, at current prices, financial transactions involving oil exceed eight billion US dollars a day.

Finally, it is important to emphasize that the levels of consumption in producer countries are growing very rapidly. Consequently, their export capacity is destined to decline. For instance, it is estimated that within ten years Russia will begin to reduce its exports of natural gas to European countries.

Table 6a Production and consumption of crude oil in thousands of barrels per day (1 barrel = 159 liters or 42 US gallons).

Country	Production	Country	Consumption
Saudi Arabia	11 161	United States	18 835
Russia	10 280	China	9 758
United States	7 841	Japan	4 418
Iran	4 321	India	3 473
China	4 090	Russia	2 961
Canada	3 522	Saudi Arabia	2 856
United Arab Emirates	3 322	Brazil	2 653
Mexico	2 938	South Korea	2 397
Kuwait	2 865	Germany	2 362
Iraq	2 798	Canada	2 293
Italy	95	Italy	1 486
WORLD	**83 576**	**WORLD**	**87 439**

Source for the three Table 6a–c: *BP Statistical Review, 2012*.
Global oil consumption (87.4 million barrels per day) exceeds global production (83.6 million barrels per day) because the figure also takes into account the "synthetic" oil produced from coal and biofuels.
Global production (about 83.6 million barrels per day) is not expected to grow significantly in the next few years. This is the main cause of the price increases since 2003. Some people take this as a sign that the peak of conventional oil production has been reached.

Table 6b Production and consumption of natural gas in billions of cubic meters per year.

Country	Production	Country	Consumption
United States	651.3	United States	690.1
Russia	607.0	Russia	424.6
Canada	160.5	Iran	153.3
Iran	151.8	China	130.7
Qatar	146.8	Japan	105.5
Norway	102.5	Canada	104.8
China	101.4	Saudi Arabia	99.2
Saudi Arabia	99.2	United Kingdom	80.2
Algeria	78.0	Germany	72.5
Indonesia	75.6		
Italy	7.7	Italy	71.3
WORLD	**3 276.2**	**WORLD**	**3 222.9**

The only energetic "macro-classification" that sees Italy among the top ten places in the world is that of gas consumption. Domestic production in Italy reached its peak in 1994 at 20 billion m^3; since then, production has fallen dramatically by almost 70%. Curiously, Iran, one of the leading world producers, can barely meet domestic demand. The United States has a huge production but not enough to cover its own consumption. Norway is among the "gas sheikhs" today with a production 26 times higher than its domestic demand and is a major exporter to continental Europe, including Italy.

Table 6c Production and consumption of coal in millions of toe per year. (Source: *BP Statistical Review*, 2012)

Country	Production	Country	Consumption
China	1 956.0	China	1 839.4
United States	556.8	United States	501.9
Australia	230.8	India	295.6
India	222.4	Japan	117.7
Indonesia	199.8	South Africa	92.9
Russia	157.3	Russia	90.9
South Africa	143.8	South Korea	79.4
Kazakhstan	58.8	Germany	77.6
Poland	56.6	Poland	59.8
Colombia	55.8	Australia	49.8
Italy	negligible	Italy	15.4
WORLD	**3 955.5**	**WORLD**	**3 724.3**

Table 6c shows an impressive consumption of coal by China, which, although is not restricted by international climate agreements to reduce its emissions of carbon dioxide, has plans to do so.

The Hidden Treasure

The ten countries that possess the largest deposits of fossil fuels are listed in Table 7a–c; Italy has been included for comparison. It is evident that the major reserves of *conventional* oil are owned by the countries of the Middle East (however, see Chapter 14), whereas the largest reserves of natural gas are found in Russia, which alone has about a quarter of the world's total supply (Table 7b).

More than 50% of the gas reserves are held by only three countries; one of these, Qatar, is smaller than the tiny state of Connecticut in the United States. Also, 40% of the world's supply of gas is extracted from fewer than 20 sites, which makes it extremely sensitive from the geopolitical point of view.

Coal reserves are distributed among a larger number of countries, and, curiously, there are no significant reserves in the Middle East, as attested by the data of Table 7c. Precise data concerning the reserves are always difficult to know accurately, not so much for technical reasons but because multinational energy companies and the producer countries often have no interest in providing accurate data. For example, in petroleum exporting countries of the OPEC Organization, production and sales quota are fixed on the basis of declared reserves. In other words, there exists an agreement among these countries in that the more oil you have (or are said to have) the more you can sell.

This could be the reason why in 2001, Qatar re-evaluated its reserves from 4 to 13 billion barrels, while Iran increased its reserves by 40% in 2004. From 2007 to 2011, the reserves of Venezuela increased from 87 to 297 billion barrels. By contrast, in 2004 Royal-Dutch Shell, one of the largest oil companies in the world, after the revelation of certain confidential documents, admitted that its stated oil reserves were 20% lower (3.9 billion barrels) than previously announced.

Table 7a Oil reserves in billions of barrels. (Source: *BP Statistical Review*, 2012)

Country	Oil reserves	Percentage
Venezuela	296.5	17.9
Saudi Arabia	265.4	16.1
Canada	175.2	10.6
Iran	151.2	9.1
Iraq	143.1	8.7
Kuwait	101.5	6.1
United Arab Emirates	97.8	5.9
Russia	88.2	5.3
Libya	47.1	2.9
Nigeria	37.2	2.3
Italy	1.4	0.1
WORLD	**1 652.6**	**100.0**

Eight of these countries belong to OPEC and hold over 80% of the world's reserves. Data include estimated reserves from oil sands in Canada and Venezuela. At present, shale sources of oil do not appear in the statistics because they are more difficult to squeeze from underground sites. Including oil from shale rocks would place the world's oil reserves at almost 5000 billion barrels (over 2000 billion in the United States alone). It should be noted, however, that the possibility of extracting large amounts of oil from this extreme unconventional source in a cost-effective manner remains to be seen.

Table 7b Reserves of natural gas in billions of cubic meters. (Source: *BP Statistical Review*, 2012)

Country	Gas reserves	Percentage
Russia	44 600	21.4
Iran	33 100	15.9
Qatar	25 000	12.0
Turkmenistan	24 300	11.7
United States	8 500	4.3
Saudi Arabia	8 200	3.9
United Arab Emirates	6 100	2.9
Venezuela	5 500	2.7
Nigeria	5 100	2.5
Algeria	4 500	2.2
Italy	100	0.05
WORLD	**208 400**	**100.0**

Nearly than 50% of the gas reserves are held by only three countries: Russia, Iran and Qatar. Almost half of it comes from less than 20 super-gigantic fields, each under the watchful eye of a handful of soldiers. The gas is therefore the most delicate fossil resource from the geopolitical point of view. Not by chance, Russia and Iran are increasingly present in newscasts.

Table 7c Coal reserves in millions of tonnes. (Source: *BP Statistical Review*, 2012)

Country	Coal reserves	Percentage
United States	237 295	27.6
Russia	157 010	18.2
China	114 500	13.3
Australia	76 400	8.9
India	60 600	7.0
Ukraine	33 873	3.9
Kazakhstan	33 600	3.9
South Africa	30 156	3.5
Colombia	6 746	0.8
Canada	6 582	0.8
Italy	10	0.001
WORLD	**860 938**	**100.0**

The first three countries in the list, which together hold approximately 60% of the world's coal reserves, are also the world's three largest consumers of primary energy (see Table 5). A massive increase in the use of this domestic resource on the part of these countries, to replace the declining availability of oil and gas, would have devastating effects on our planet's climatic stability.

In the months following this revelation, Shell cut its reserves by another 6 billion barrels.

Energy Also Travels

To add complexity to the problems of the use of fossil fuels, 60% of the world's production is destined to consumer countries far from those producer countries where deposits are located.

Where the land topography permits, a dense network of pipelines extends from the producer countries to the consumer countries. Added to those pipelines already in existence running through the length and breadth of Italy, the coming years will see a 900-km pipeline, which will bring Algerian natural gas to Tuscany after having crossed the entire island of Sardinia. Many of the oil pipelines in use today are over twenty years old, an age that does not make them completely safe. Particularly dangerous is the situation in Alaska, where global warming is melting the permafrost through which runs one of the principal oil pipelines in the world.

For longer distances, fossil fuels are transported by sea by oil tankers or by Liquefied Natural Gas (LNG) tankers. This entails a strong energy flow density in sensitive geographical areas, such as the Strait of Hormuz, an obligatory passage point of the oil route in the Persian Gulf through which passes a quantity of crude oil equal to the annual consumption of the whole of Europe, or the Strait of Malacca, through which 25% of the oil is transported by sea, in large part intended for China, Japan, and South Korea.

Sea supertankers that transport crude oil and which can carry a load exceeding 300 000 tons are very expensive to operate because they must conform to strict rules to avoid catastrophic environmental accidents such as those that have occurred in the past (e.g., along the Alaskan coast line – remember the Exxon Valdez?).

Insofar as natural gas is concerned, the gas must first be liquefied at a temperature of −162 °C at the port of departure for transport by LNG tankers and later re-gasified in the port of arrival. These operations are technically complex, costly from the economic and energy points of view, and potentially dangerous.

The United States is geographically distant from the major producers of the Middle East, and so must be supplied by sea. Accordingly, they are turning their attention increasingly toward neighboring countries on the American continent – Venezuela and Bolivia – where the supply of oil and gas to the United States is one of the most contested political issues.

Europe has very little fossil fuel reserves, but from the viewpoint of supply finds itself in a much more favorable geographical situation than the USA, as it is virtually surrounded by the principal producing basins of the world (Russia, the Middle East, North Africa, the North Sea).

It is essential to own the deposits, but it is also important to control the routes through which fossil fuels are distributed. We know this full well in Europe, where the flow of gas from Russia and Libya has already suffered outages for political and military reasons. Accordingly, the energy highways are often chosen based on political criteria rather than the minimum distance.

Costly Energy Invoices

The consumption of fossil fuels has reached high levels (Table 6a–c), and certainly the demand for this convenient and powerful source of energy will continue to increase in the coming years.

According to market rules, as the demand for oil increases so will the price of the commodity, which exceeded 50 US dollars a barrel in the Fall of 2004 – in July 2008 the price of a barrel was up to $150 but declined to around US $100 in August 2012. In any case, the consumer is faced with huge daily fluctuations at the pump.

The above notwithstanding, the price of oil is expected to increase further because of other factors. The first investments in drilling for oil in the early 1900s produced significant yields because it was not necessary to dig deep wells – oil flowed easily. With the passage of time, research and exploitation of new deposits has led to increased costs. At the same time, the percentage of oil remaining in the wells is of the worst quality possible.

Another negative factor is the concentration of deposits in a few countries, many of which are politically unstable. Recent wars and international terrorism, in part generated and nurtured by inequalities and unfair distribution of the Earth's resources, are causing further doubts about the possibility of arranging and ensuring with any certainty the continuity of oil resources. Therefore, every country tends to accumulate its own reserves as much as it can. Oil has become a refuge for investors, as gold is, so much so that major international banks now invest in oil. This creates private oil reserves.

Politicians tend to ignore, downplay, and even discount the problem of further increases in oil prices. They think that it would be counter-productive for their re-election if they had to give unfavorable news to the citizens and explain the root causes of this ever-increasing trend of cost increases. Some pundits believe that this may have political consequences in the race for the United States Presidency in 2012.

It should be added that the price of oil, and energy in general, has little to do with actual production costs. The energy industry has for a long time received generous public subsidies and resents economic agreements that have little to do with technical costs. Added to this is the colossal expense that many countries, particularly the richest ones, sustain so as to maintain the continuous flow of oil, gas, and coal for various purposes: for example, military deployments, funding to friendly and moderate countries, and interference in the internal politics of producing States, and so on and so on.

The military apparatus deployed in the Persian Gulf, in the Indian Ocean, and in the Mediterranean Sea to guard the energy corridor of the Middle East comes at an astronomical cost. The costs multiply manifold when this military machine is in action. According to official sources of the United States, by 2017, spending for the three *oil wars* (two in Iraq and one in Afghanistan) will amount to 2400 billion US dollars, equivalent to about 8000 dollars for every American citizen.

The above considerations give you an idea that no one in the world knows how to estimate accurately the real price of a barrel of oil.

Alliances, Tensions, Wars

It is clear that, in such a situation, the rush to energy supplies strongly influences both the policies of the producing states, which have an interest in selling, and the policies of the consumer states, where the economy and social stability would collapse without the availability of fossil fuels. Here's the reason for strategic military alliances, such as the one between the United States and Saudi Arabia, and the pursuit of energy supplies by many other consumer countries, notably China and India, which concluded bilateral agreements with some producing countries such as Angola. And here is also the reason for strong tensions that can lead to wars.

Frankly speaking, it's hard to believe that the wars in Libya and in Iraq had nothing to do with the control of major producer countries of oil and gas and the control of related land and sea routes through which energy resources pass (Afghanistan, Chechnya, and Serbia were inhospitable and poor in resources.) In the United States, which imports huge amounts of oil from Arab countries, many motorists when filling up at the pump have the feeling of financing both sides simultaneously in the fight against terrorism.

In the wars for oil, the environment has suffered considerably. During the 1991 Gulf war, 730 oil wells were blown up, many of which burned for several months with a loss of about 240 billion liters of crude oil (1.5 billion barrels), equivalent to 2% of the reserves of Kuwait. The production of air pollutants from this uncontrolled oil combustion was not less than catastrophic for the environment. And during the same war, 1.7 billion liters (about 11 million barrels) of oil ended up in the Persian Gulf.

5
Collateral Damage

You will never have enough of what you did not need to be happy.

Eric Hoffer

It is known, though too often many seem to forget, that every time a resource is used, wastes are inevitably produced. Wastes are never *innocent*. In one way or another, and sometimes in many ways, such wastes are harmful.

In recent decades we have come to realize, with even greater concern, that the use of fossil fuels produces gaseous toxic substances that are harmful to human health, to the environment, and to climate stability.

The effect of contaminants is not confined to the places where they are produced, as they diffuse on a planetary scale, thus making the phrase *not in my back yard* (NIMBY) total nonsense. The NIMBY syndrome is often articulated by individuals, by communities, and by countries in the hope of avoiding harm to their own territory.

Slowly, however, awareness is spreading that the environment is a common good and that therefore its preservation requires actions both at the local level and globally. Appropriate measures need to be taken in the immediate future and for the long term.

The history of the last decades has shown that serious harm can be caused to humans and to the environment by unexpected phenomena and by less important processes. A classic example is the *hole in the ozone layer* caused by emissions of chlorofluorocarbons–chemicals whose use is limited but which are nonetheless capable of destroying the ozone layer in the stratosphere. The ozone layer protects people from the high-energy solar ultraviolet radiation (wavelengths below 290 nm). Accordingly, we should use extreme caution in introducing any waste materials into the air, the waters, and the soil. There is a need for widespread and continuous monitoring of the situation, together with a thorough scientific analysis of any data collected.

Under the influence of public opinion and thanks to technological progress, developed nations have begun to reduce emissions of harmful substances. But they now want to impose their ecological standards on developing countries, which can ill afford the related costs necessary to control and limit pollution, lest their products would never make it to the market place.

Powering Planet Earth: Energy Solutions for the Future, First Edition. Nicola Armaroli, Vincenzo Balzani, and Nick Serpone.
© 2013 Wiley-VCH Verlag GmbH & Co. KGaA. Published 2013 by Wiley-VCH Verlag GmbH & Co. KGaA.

Pollution of the planet is expected to increase, partly because developed nations refuse to adhere to strict pollution standards so as not to affect adversely their own industries, as they have to compete with the rest of the world for market share.

The Planet Overheats

Nearly a quarter of the solar radiation hitting the Earth is reflected back into space by the clouds; about another quarter is absorbed by the atmosphere and is transformed into heat. The remaining half of the solar energy reaches the Earth's surface; of this fraction, 90% of it is absorbed and heats the Earth, while the remaining 10% is reflected in part by the polar ice caps.

Some of the energy absorbed by the Earth is radiated back into the atmosphere and absorbed by gases (e.g., water vapor [H_2O], methane [CH_4] and carbon dioxide [CO_2]), following which the heated gases transmit the infrared radiation (i.e., heat) back to Earth. In practice, these gases function as a greenhouse: they allow sunlight to enter, but prevent the resulting heat from escaping into the stratosphere (the greenhouse effect). If there were no natural greenhouse gases, the temperature at the Earth's surface would be around 30 °C than what we experience daily. Thus, natural greenhouse gases do have a positive effect on climate. However, ever since mankind began to make extensive use of fossil fuels, increasing quantities of greenhouse gases generated by his activities, particularly CO_2, are emitted continuously into the atmosphere, and so bring about significant changes to Earth's climate. Let's now see how this might happen.

Fossil fuels are formed in the hold of spaceship Earth in the absence of oxygen through the transformation of organic plants and animal matter through very complex chemical processes that have taken place in the course of hundreds of millions of years. Therefore, fossil fuels are, in a sense, solar energy stored in the form of chemical bonds between carbon atoms (C–C) and between atoms of carbon and hydrogen (C–H).

When we extract fossil fuels and allow them to react (burn) with oxygen of the air, the C–C and C–H chemical bonds break down and form other types of chemical bonds, for example, C–O and H–O bonds. These chemical transformations release large amounts of energy together with the formation of carbon dioxide and water. For instance, 1 g of coal develops an equivalent quantity of heat equal to 32.8 kJ and produces 3.66 g of carbon dioxide (mass has increased because carbon has combined with air oxygen).

Oil and natural gas are mixtures of hydrocarbons – these are compounds that contain carbon and hydrogen only. Hydrocarbons may come in the form of gases, liquids, or solids, depending on the number of carbon atoms in their respective molecules. Thus, gasoline obtained from the distillation of petroleum is a mixture of liquid hydrocarbons (approximate formula: C_8H_{18}), which react with oxygen yielding water and carbon dioxide: 1 g of gasoline (petrol) develops a quantity of heat equal to about 47.8 kJ and produces about 3.08 g CO_2. Natural gas is composed mainly of methane, CH_4, which reacts with oxygen to produce water and carbon dioxide: 1 g of methane develops 55.6 kJ in the form of heat and produces 2.74 g of carbon dioxide. Accordingly, when we use coal, oil, and methane to extract the stored energy, they produce a quantity of CO_2 equal to about three times their weight.

At the current pace, every year mankind produces and emits about 30 billion tons of carbon dioxide into the atmosphere. The main producers of this waste gas are the United States, Europe, Japan and China. Each year an American produces about 18 tons of CO_2, nearly triple the amount produced by an Italian, who in turn produces five times more than an Indian. The amount of CO_2 that enters into the atmosphere from Italy is about 70 times greater than the quantity of CO_2 from Ethiopia whose population is comparable to that of Italy.

Following the use of fossil fuels from the beginning of the industrial revolution to the present, the concentration of CO_2 in the atmosphere has increased from approximately 275 to about 400 parts per million (ppm). If no appropriate action is taken, the quantity of CO_2 is estimated to increase to about 550 ppm by the end of the twenty-first century. The consequences of such an increase in CO_2 concentration in the atmosphere–which someone has defined as *a dangerous experiment out of control*–could prove disastrous. Moreover, doubling the concentration of CO_2 could cause an increase in global average temperature of about 3 °C, accompanied by a rise in sea levels and increased frequency of extreme weather: for example, heat waves (drought) and heavy precipitations (floods, mud slides).

Compounding this harsh reality is the fact that poor countries are most vulnerable to climate changes. This is demonstrated by the number of victims and the extent of damage caused by hurricanes that hit the Caribbean (e.g., Haiti) and the United States.

Agreements and Disagreements

The first scientific data on the increase in the concentration of CO_2 in the atmosphere can be traced back to 1957. Only after 31 years had elapsed did the United Nations address this problem with the creation of a special Commission to examine climate change: the *First International Panel on Climate Change* (IPCC).

At the 1992 Conference held in Rio de Janeiro (Brazil), the IPCC produced an agreement (Protocol) that included specific actions to reduce greenhouse gas emissions. This Protocol was then amended and approved in Kyoto, Japan, in 1997–referred to as the Kyoto Protocol. The acceding nations pledged to reduce their greenhouse gas emissions by between 5 and 10% by 2012 relative to 1990 emission levels. The agreement, however, remained dormant for a long time owing to the non-ratification of the Protocol by the United States, Australia and Russia. The protocol came into force following its ratification by Russia on September 30, 2004. Unfortunately, the Kyoto Protocol has not had a big impact–in

Figure 8 Cover of the magazine *Science* of March 24, 2006, inviting people to *break the ice* and accept the available evidence of global warming caused by our massive use of combustible fossil fuels.

part due to the non-adherence of the United States, by far the largest producer of CO_2, and because the expected reductions were modest at best.

The last few years have seen, particularly in the United States, intense action taken by some scientific institutes, financially supported by the oil industry, to shed doubt on the notion of global warming. However, the IPCC report of February 2007, confirmed the rise in the Earth's surface temperature, the melting of glaciers, and the rise of sea level. The report also stated that all this was caused – with a probability greater than 90% – by human activities.

The IPCC was awarded the 2007 Nobel Peace Prize for its valuable activities, together with former American Vice-President Al Gore, author of the documentary film *An Inconvenient Truth* dedicated to climate change.

Despite the large diplomatic efforts, the UN Conferences on climate change held in Bali in 2007 and then in Copenhagen in 2009 led only to a general commitment to limit global heating, containing it to within 2 degrees Celsius. At the Rio+20 (Earth Summit) conference of June 2012, a document was approved aimed at guiding the world toward a more sustainable future and all nations "reaffirmed" their commitments to phase out harmful fossil fuel subsidies.

Jailing the Offender?

About 30% of the emissions of the greenhouse gas carbon dioxide originate from the use of coal. There are currently 2300 coal-fired power plants in the world – one is being commissioned every week in China. From these facts alone, and considering that coal is the most abundant fossil fuel, there is a movement afoot that proposes continuing to use this fuel in power plants that would not emit carbon dioxide. To accomplish this feat, however, requires that the carbon dioxide produced by burning coal be sequestered *before* it is emitted into the atmosphere. At least a dozen different techniques have been proposed for this purpose. The most promising of these is the capture of CO_2 and its storage underground in spent oil wells and natural gas caves. Theoretical assessments are seemingly comforting, and ongoing searches seem to give satisfactory results. However, such assessment studies are only now moving from laboratory phase to that of pilot plant scale.

From a technical point of view, this methodology would isolate the CO_2 from other exhaust gases, and would then compress it and transport it through a pipeline to the indicated storage area. Various solutions have been proposed for the first stage of the process, the most complicated one. Apparently, with appropriate modifications, old power plants could be converted to store CO_2. There are people who argue, however, that these modifications would cost more than a new power plant.

Regardless, the capture of carbon dioxide will surely increase the cost of electricity produced from coal to the end user – estimates of such increases vary from 20 to 80%. Another critical aspect of these technologies is the concrete possibility of escape of CO_2 into the atmosphere through various paths that are not, *a priori*, easily predictable.

Finally, the possibility of incidents should not be underestimated. For example, the escape of large amounts of CO_2 from storage, a gas heavier than oxygen, could cause serious damage to the population, as was the case following the 1986 eruption of 80 million cubic meters of CO_2 from Lake Nyos in Cameroon, killing 1800 people.

With regard to the investments in the above-mentioned technology, it's worth reflecting on what John Turner of the National Renewable Energy Laboratory in the United States stated – and we quote:

> That large amounts of money, intelligence and energy should be invested in technologies for the sequestration of CO_2 remains an open question: is this the best way to spend our limited financial and energy resources? The mere fact of possessing large reserves of coal does not mean that we must necessarily use them. The point is whether we should leave those resources underground or move decisively towards something more advanced.

A Subtle Danger

In addition to carbon dioxide, the use of fossil fuels also emits a large number of other substances into the atmosphere that are harmful to health. This happens for two reasons.

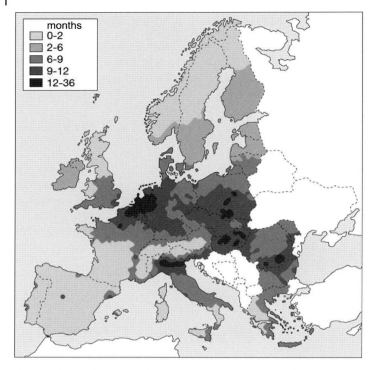

Figure 9 Atmospheric particulate matter produced by the use of fossil fuels is harmful to health. This map of Europe indicates a reduction in the average life expectancy (in months) resulting from exposure to PM-2.5. Data source: European Commission, DG Environment, 2005.

First of all, fossil fuels found in deposits are always mixed with more or less relevant quantities of substances of all kinds – sulfur compounds, heavy metals, hydrocarbons, and aromatic compounds, which are only partially separated from the fuel prior to its use. Almost all of these substances and their products present in the form of atmospheric particulates are dangerous to human health. In general, coal is dirtier than oil, which in turn is dirtier than natural gas. The chemical composition and amount of impurities contained in fossil fuels vary depending on the source of the deposits.

Secondly, combustion processes that take place in injection engines and in steam boilers do not use pure oxygen as the oxidizer, but use air at the high temperatures generated by the combustion process. This causes nitrogen and oxygen of the air to combine and produce nitrogen oxides – the so-called NOx gases. These are inherently polluting gases and precursor chemicals of other harmful substances such as, for example, ozone and atmospheric particulates.

The current focus of scientific research is directed at the adverse effects of particulate matter in the atmosphere. This pollutant, of variable chemical composition, is more dangerous than smaller particles. Those with a diameter between 10 and 2.5 µm (PM-10) penetrate the **bronchi** and **bronchioles**, whereas the smaller particles (PM-2.5) penetrate the **alveoli** and can enter directly into the blood stream.

There is strong evidence of the effects of exposure to air particulates on the respiratory and cardiovascular apparatus (Figure 9). Mutagenic effects have been observed in rats. The constant exposure to particulates leads to a state of chronic inflammation. The inflammatory process of tissues produces an environment favorable to carcinogenesis.

In addition to being harmful to health, oxides of nitrogen and sulfur can interact with other chemical compounds present in the atmosphere, transforming them into fine and ultrafine dust particles (secondary particulate matter).

Ozone is another secondary pollutant produced in the atmosphere from the action of light on primary pollutants. Its presence in the air is extremely harmful to animals and plants. It's also worth noting that heavy metals can also cause carcinogenic, mutagenic, and teratogenic effects.

Rain Is No Longer What It Used to be

Beginning in the late 1960s, we began to notice that in certain regions of the globe, particularly in heavily industrialized areas, rainfall and fog had a high degree of acidity. This phenomenon is due to the presence of strong acids (e.g., nitric acid, HNO_3, and sulfuric acid, H_2SO_4) in the atmosphere.

Nitric acid is generated directly or indirectly from the reaction of water (rain, fog) with nitrogen oxides (NOx gases). Catalytic converters installed in cars transform these NOx substances into molecular nitrogen (N_2) and molecular oxygen (O_2), and thus limit the release of these NOx into the atmosphere. This technology, along with new engines, has led to an impressive reduction of NOx emissions from motor vehicles. It is also possible to equip thermoelectric power stations with catalytic converters. This practice is not yet sufficiently widespread, at least in many countries.

For now this chemical reduction process can be achieved only with platinum and rhodium compounds as the catalysts. The availability of these rare and precious metals is in danger of declining owing to the continuous increase in car production. In July 2012, the price of gold reached a value of 1600 US dollars per ounce, which caused some concern. At the same time, the price of rhodium was around $1200 US dollars per ounce.

Sulfuric acid is generated from sulfur present as an impurity in all fossil fuels. Reacting with the oxygen of the air, sulfur gets oxidized to sulfur dioxide (SO_2) and later to sulfur trioxide (SO_3), which eventually combines with water to form sulfuric acid. This acid is especially harmful because it extracts calcium ions (Ca^{2+}) from the soil–essential components for plant growth–to produce insoluble calcium sulfate ($CaSO_4$).

Acid rain changes profoundly the chemical properties of soil and fresh waters, and can even cause serious damage to plants (e.g., maple trees in North America, a source of maple sirup). In some cases, acid rain can cause total deforestation of whole areas and complete disappearance of life in lake waters. Acidic precipitations are also damaging to the integrity of monuments and buildings, especially those that are built with high content of calcium carbonate: for example, marble and travertine.

In the last twenty years, the problem of acid rain has been greatly reduced in Western countries and in Japan, but it's getting worse in China and in several developing countries.

Financial Compensation

An externality in the field of Economics is defined as a consequence of an economic activity which affects other parties without this being reflected in market prices. In the present context, it reflects a situation in which costs or benefits of the private use of goods or services differ from the costs or benefits to the community. A classic example of externality is pollution produced by a small number of people, but which harms a far greater number of individuals.

The damage that the use of fossil fuels causes to human health or to the environment is therefore an externality whose costs are generally not taken into account by producers and consumers of energy. These costs usually fall on the poor taxpayers.

In recent years, some national and international agencies have begun to assess the extent of economic damage, even though the treatment of the problem is extremely complex – there is no general consensus on the parameters with which to base the assessments. For example, a gas-fired power station with a combined power of 800 MW, one of the less polluting ways to produce electricity, generates annually 1600 tons of NOx gases and 60 tons of PM-2.5 particulates (a conservative estimate).

Health costs to the society, linked to the treatment needed by the population concerned, are estimated at 12 million Euros per year. This figure should be compared to a compensatory contribution that the owner of the power plant must pay in Italy to local administrations concerned: 0.2 Euros per MWh, amounting to 1.4 million Euros per year – note that this contribution is not intended to reimburse health costs, but is reimbursement for *the absence of alternative use of the territory and for the logistical impact of the construction sites.*

In any case, the energy utility pays for only about a tenth of the damage done to the community. This financial compensation that local administrators consider a big deal won't benefit the community; it will likely be used by elected officials to defray the cost of a small urban or cultural project: for example a paved road or a song festival.

A much more rigorous approach to so-called compensation mechanisms is possible. In the United States, especially in California, there is advanced legislation that sets a cap on greenhouse gas emissions.

Air pollution affects severely congested mega cities in poor countries, where millions of people take refuge in search of an unlikely fortune. But no less disturbing is the situation in the most remote rural areas where, in insufficiently ventilated environments, food is cooked using biomass and rudimentary furnaces that produce high concentrations of particulate matter and carcinogens. Current estimates show that in poor countries 2 to 4 million children die every year from air pollution that exists between the four walls of their habitat.

In conclusion, limitation of the use of fossil fuels is necessary to safeguard human health and the integrity of the Earth's biosphere. This is a far more serious problem than are progressive fuel shortages.

Minimize! Save the Planet

At the moment, an American consumes as much energy as 2 Europeans, 4 Chinese, 14 Indians or 240 Ethiopians. This inequality is compounded by the fact that countries that consume the least are those that are the most populated. According to United Nations forecasts, an additional 2.5 billion people will need access to energy supplies in the next thirty years.

In this situation, it is clear that it will not be possible for all the inhabitants of the Earth to live the *American way of life* using fossil fuels. There are simply not enough fossil fuels! And in a sense, we should add *fortunately* because, if there were, their widespread use would bring about additional devastating changes to the climate and cause undue harm to human health.

To ensure a future for humanity, it is therefore necessary to withdraw – albeit progressively – from the use of fossil fuels and seek alternative energy sources. As we shall see later in this book, possible solutions are basically two: (i) solar energy and renewable sources, and (ii) nuclear energy. In the next couple of chapters we will discuss in some detail the advantages and disadvantages of these two forms of energy.

The advent of a future energy crisis is unquestionable, although many people have not yet perceived its severity. This bleak assessment is based on three incontrovertible data: (i) the progressive depletion of fossil fuels, (ii) damage done to health and to the environment by the massive use of such fuels, and (iii) the huge inequality that exists between rich and poor countries in the availability of energy.

The energy crisis calls into question the growth model, based on consumption at all costs, that the great availability of energy at ridiculous prices has created in the past decades and that has benefited only a fraction of the Earth's population. In the next few chapters, we will examine some possible solutions to this crisis. We shall also see that the explosion of the energy crisis leads to other no less important problems of a cultural, ethical, and social nature.

6
Energy from the Atom

Nuclear power is as safe as a chocolate factory.

The Economist, March 29, 1986
(4 weeks prior to the Chernobyl disaster)

The atom is the smallest part of every existing element in nature. It consists of two distinct areas: a central core – the nucleus – and a surrounding area – the electronic cloud.

The nucleus consists of positively charged particles (the protons) and neutral particles (the neutrons), all in close contact with each other. The periphery of the atom is formed of negatively charged particles (the electrons) distant from each other and frantically moving around the nucleus (see Figure 10).

Scientists who first discovered the structure of the atom at the beginning of the twentieth century (note that the Greek philosopher Democritus had also speculated about the existence of atoms 2300 years ago) were dismayed to find that the volume of the atom is defined essentially by the electronic cloud, whereas the mass is almost exclusively concentrated in the nucleus. In other words, the atom consists of a very large but otherwise evanescent volume/space and a very small but otherwise heavy nucleus. Accordingly, atoms are rather eccentric objects that even a very imaginative architect would not have designed that way.

Combustion processes that provide the vast majority of the world's energy involve only the peripheral components of the atom, namely the electrons. Chemical bonds break and form through perturbations of the electronic cloud of the atoms involved. During this hectic activity, atomic nuclei remain unchanged and everything that happens around does not concern them. After 70 million years of deep sleep in underground sites, during the combustion of methane (CH_4), the space surrounding the nucleus of the carbon atom is transformed from a tetrahedral electron cloud in which the C atom is bound to 4 hydrogen atoms to a linear form in which the C atom is bound to 2 oxygen atoms – carbon dioxide (CO_2).

Since the discovery of fire and for thousands of years afterwards, things have moved forward in this way, without any variations on the theme. All the processes of transformation of matter developed by man have involved only the electronic cloud of atoms, leaving the nucleus unchanged. However, things have changed since the end of the nineteenth century, when scientists gradually began to unravel

Powering Planet Earth: Energy Solutions for the Future, First Edition. Nicola Armaroli, Vincenzo Balzani, and Nick Serpone.
© 2013 Wiley-VCH Verlag GmbH & Co. KGaA. Published 2013 by Wiley-VCH Verlag GmbH & Co. KGaA.

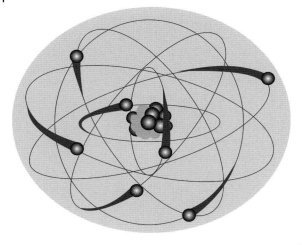

Figure 10 Simplistic outline of the structure of the atom. Note that the figure is not to scale. In fact, the volume occupied by the nucleus is 100 000 times smaller than the electrons' orbits. Illustration courtesy of Charlotte Erpenbeck-Shutterstock.

the structure of the atom and the role of the nucleus. This was thanks to the work of such scientists as Wilhelm Röntgen, Henry Becquerel, Joseph Henry Thomson, Marie and Pierre Curie, Ernest Rutherford, James Chadwick, and Enrico Fermi who led the famous *boys of Panisperna Street* group at the University of Rome (La Sapienza) that included the well-known physicists: Amaldi, D'Agostino, Majorana, Pontecorvo, Rasetti, and Segrè.

The die was definitively cast on December 2, 1942, when – in a somewhat unusual place, a gymnasium of the University of Chicago – Fermi showed that it was possible to produce (nuclear) energy in a controlled manner simply by *poking* at the atomic nuclei.

With the *atomic pile*, Fermi had lit the nuclear fire whose power was immensely superior to that of traditional fire – it opened up a new era. Changed forever were the ways we see the world and the way we manage and handle international relations. Born was the tangle of hopes, fears, questions, rivalry, tensions, and wars that we continue to debate to this day. From that day, Spaceship Earth became more fragile.

Splitting the Atom

To perform that historic experiment, Fermi had not chosen atoms at random. He used the heaviest chemical element found in nature, Uranium (U), which consists of 99.3% of atoms containing 92 protons and 146 neutrons in the nucleus. The sum of these two numbers, 238, represents the atomic mass of uranium. It is therefore called *uranium-238*, or simply ^{238}U.

The uranium needed for the production of nuclear energy, however, is the small fraction that remains, namely the 0.7% of the total. This fraction is made of an *isotope*, a slightly different type of uranium, which in the nucleus also contains 92 protons but only 143 neutrons; this is uranium-235 or ^{235}U.

Only ^{235}U is *fissionable;* that is, it is unstable and apt to produce energy by splitting its nucleus. In other words, the ^{235}U nucleus can be "broken," producing nuclei of lighter chemical elements and releasing a large amount of energy (namely, electromagnetic radiation and kinetic energy of particles produced by the nuclear reaction). In this regard, Uranium-235 is the only existing fissionable atom available in appreciable amounts in nature. This isotope of a chemical element is uncommon, and represents a primary energy source, just like the Sun and crude oil.

The fission reaction of ^{235}U is triggered by the absorption of low-energy neutrons (*n*), creating two smaller nuclei and releasing 2 or 3 neutrons. For example, ^{235}U can be split to yield a barium atom (^{141}Ba) and one atom of krypton (^{92}Kr), plus three times the number of neutrons that had been used to initiate the nuclear process – as shown in Figure 11 – and the release of lots of energy.

Fission causes the transformation of a small amount of mass into a large amount of energy according to Einstein's law $E = mc^2$: every gram of ^{235}U that undergoes fission releases about 84 MJ in the form of high frequency electromagnetic radiation, equal to the energy obtained by burning about 20 tons of oil.

A single nuclear fission releases 2 or 3 neutrons, which – in the presence of an appropriate *moderator,* that is, a substance capable of slowing down the neutrons,

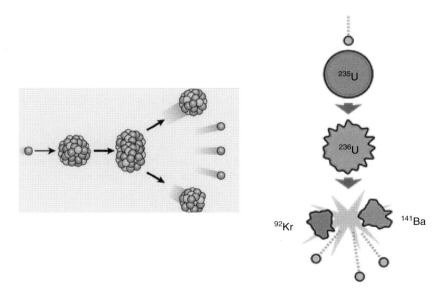

Figure 11   Scheme illustrating the fission of the core of one atom of ^{235}U, triggered by a slow neutron, with the formation of a nucleus of ^{141}Ba and one of ^{92}Kr. Fission produces three neutrons, which (if slow) can split other uranium nuclei, thus causing the chain reaction.

such as water–can in turn cause the fission of other nuclei of ^{235}U through a chain reaction. For this to occur, however, it is necessary that at least one of the *new* neutrons does not escape from the system before causing another fission reaction. This condition occurs when the mass of the fissionable material exceeds a particular value, the so-called *critical mass*–at that point, the chain reaction becomes self-sustaining.

Without appropriate control, the number of active neutrons multiplies during the individual events of the fission reaction. In this case, the speed of the process increases rapidly causing an explosion, as occurs in an atomic bomb. If, however, we introduce *control bars* into the mass of the fissionable material that are made of materials that absorb the neutrons–such as cadmium or boron–then we can control the speed of the process, prevent the explosion from occurring, and allow us to harness the heat developed when the neutrons and radiation collide with the walls of the container. This is what happens in nuclear power plants for the production of electricity.

We have seen in Figure 11 a nuclear reaction that produces barium and krypton. However, fission of ^{235}U can also take place through other pathways to produce many other types of lighter isotopes of other elements, with a number of protons between 30 (zinc, Zn) and 65 (terbium, Tb). Overall, they comprise over one hundred different isotopes. Among these are various radioactive elements that are extremely harmful to health, namely, cesium-137 (^{137}Cs), iodine-131 (^{131}I) and strontium-90 (^{90}Sr). These became well known to the general public in the spring of 1986 following the explosion of one of the reactors of the Chernobyl nuclear complex in the Ukraine. The fallout of these and many other toxic and radioactive elements was felt throughout Europe.

The radioactivity of atoms decreases with the passage of time. Every isotope has a *half-life,* which refers to the time it takes for half of the nuclei of a given radioactive sample to be transformed into nuclei of other isotopes or other elements. The radioactivity of the iodine-131 isotope (^{131}I) disappears within a few days, while the half-life of cesium-137 (^{137}Cs) and strontium-90 (^{90}Sr) is about 30 years, making these two radioactive isotopes hazardous for centuries.

Insofar as the longevity of the isotopes of the starting material, ^{235}U and ^{238}U, is concerned, they have a half-life of 704 million years and 4.5 billion years, respectively. Accordingly, there are no immediate concerns regarding the expiry date of the natural supplies of uranium on Earth.

Nuclear Accidents

Prior to the Fukushima disaster of 2011 in Japan, which will be discussed in some detail in Chapter 9, there had been other no less serious accidents in nuclear power plants. Figure 12 illustrates the nuclear fuel process.

On April 26, 1986, reactor number 4 of the Chernobyl nuclear complex in the Ukraine, then part of the Soviet Union, went out of control and exploded releasing into the atmosphere. within a matter of about 10 days, 6.7 tonnes of radioactive

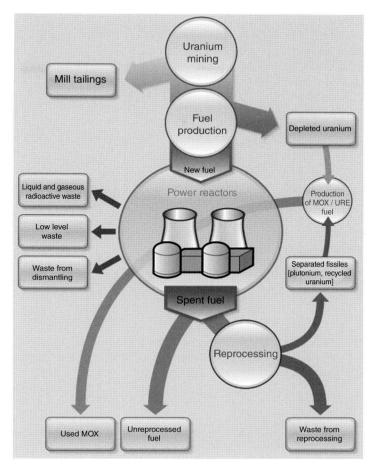

Figure 12 The nuclear fuel process displaying the various stages from mining uranium to ultimate waste. Source: http://en.wikipedia.org/wiki/Nuclear_fuel.

material that contaminated not only the vast areas close to this nuclear installation, but also, albeit less so, some areas of Eastern Europe, Scandinavia, and marginally some areas of Western Europe.

It's been estimated that the incident affected – more or less directly – about 8.4 million people and put out of production some 784 000 hectares of agricultural land, and 694 000 hectares of forests. To this day an area about 30 km² around the Chernobyl plant remains highly contaminated. There are no plans in place to secure the thousands of tons of radioactive material produced by the destroyed reactor. There is also the danger of collapse of the provisional concrete sarcophagus used to cover the reactor.

After 25 years, despite various reports prepared by International Organizations, it is difficult to take stock of the disaster because the radiation effects will be felt for a very long time. About 600 000 people, comprising the reactor staff, residents, emergency teams, and rescue workers, were affected by the emitted radiation. The immediate victims were 56, although other sources speak of 200. According to recent estimates, to these should be added an unspecified – but very high – number of people who have died prematurely because of exposure to high doses of radiation. Approximately 4000 Ukrainian children have been affected by thyroid cancer caused by radioactive iodine, 15 of whom died. Fortunately, others have been cured.

Serious, and initially underestimated, has been the psychological damage to the population: 350 000 people were evacuated, 116 000 of whom immediately after the incident. Only some have been able to return to their homes. Many have problems of mental balance. They no longer have confidence in their state of health, have serious problems with depression, lack the capacity for initiatives, and have often fallen into a paralyzing fatalism. Alcoholism is widespread. There are also demographic problems as the more educated young people have left the area for economic reasons, since everything that comes from the Chernobyl area is viewed with suspicion.

To secure the remains of the Chernobyl reactor for the next one hundred years necessitates the construction of a gigantic arched structure (NSF, *new safe confinement*) 110 meters high and 270 meters wide, which will be placed in position through the use of a huge rail system. The estimated cost of this project is 1.5 billion Euros, allocated mainly by the European Union. The overall costs of the disaster will never be assessed with any precision. Nonetheless, it is estimated to be of the order of hundreds of billions of US dollars.

Several other serious accidents have been near misses in the last 50 years. The most serious in the history of nuclear power in the United States occurred in 1979, when the Three Mile Island reactor in Pennsylvania reached near complete fusion of the core. In France, an incident with a partial core meltdown occurred in 1980 at the Saint-Laurent nuclear installation. Yet another very serious event occurred in 2002 at the Davis-Besse plant in Ohio. As a result of the lack of safety inspections, a loss of water rich in boric acid gradually corroded the reactor's 15 cm thick stainless steel cover to near perforation. If the problem had not been discovered in time, the consequences would have been disastrous.

The earthquake measuring 6.7 on the Richter scale that hit central Japan on July 16, 2007, caused 11 casualties and many wounded. The event affected the largest nuclear power installation in the world, namely the Kashiwazaki-Kariwa nuclear reactor located only 20 miles from the epicenter. As a result of the earthquake, an ensuing fire to a transformer triggered a chain of events that damaged a number of barrels of radioactive waste. This led to the release of an unknown amount of radioactive materials, including liquids, into the environment.

That power plant, like the Fukushima nuclear complex, was operated by the Tokyo Electric Power Company (Tepco), a giant of the industry which has been at the focus of repeated scandals for falsifying data on controls and safety of its facilities. The Japanese authorities closed down the plant for 21 months and forced the introduction of higher safety standards than when it was designed over 30 years ago. As we have now come to realize after Fukushima, however, these unfortunate events have failed to sound the alarm with sufficient urgency to prevent future nuclear disasters.

An Inconvenient Legacy

We saw earlier that fission reactions of ^{235}U in nuclear reactors occur in the presence of a much greater number of atoms of the more abundant ^{238}U isotope. A small fraction of this natural uranium also absorbs neutrons and is transformed into ^{239}U, which quickly decays into Plutonium-239 (^{239}Pu).

Plutonium is practically absent in nature and can only be obtained through nuclear reactions. It is so toxic that inhalation of just less than a millionth of a gram ($<10^{-6}$g) suffices to develop lung cancer. Its radioactivity is practically ever-lasting on human timescales, requiring 24 000 years for its concentration to decrease by a half. On August 9, 1945, a bomb containing 6 kilograms of plutonium razed the city of Nagasaki, causing 80 000 casualties instantly. Many thousands died or became ill in the following decades because of the devastating effects of radiation.

Production of electricity via the nuclear route means a need for *factories* to produce a material – *par excellence*, plutonium – suitable to build nuclear weapons. However, this material must be extracted from all the remaining mass of exhausted fuel by the technique called *reprocessing*, that requires very advanced technologies. In 1977, in an attempt to set a good example, the then American President Jimmy Carter forbade the reprocessing of spent nuclear fuel in the United States, however, no other nations followed Carter's decision. As described in Chapter 9, countries that possess the technology to run reprocessing at the moment are France and the United Kingdom. Of course, the President's decision did not prevent the United States from continuing building its atomic warheads by other means.

Plutonium is probably the most dangerous substance that man has ever created. From the beginning of the atomic era, nuclear power stations have produced approximately 1500 tonnes of plutonium, that is, 15 million billions of carcinogenic doses or more than two million doses for each inhabitant of planet Earth. Despite the enormous amount of money invested to secure this toxic legacy, no real solution has yet been found and probably never will be.

Where Do We Store Nuclear Wastes?

Nuclear residues or wastes are divided into two categories: (i) *low and medium radioactive wastes*, including the equipment used for processing the fuel, contaminated soils, pieces of dismantled equipment, and protective devices for the staff of the nuclear reactors; and (ii) exhausted or reprocessed fuel, which constitutes *highly radioactive* waste, and must be kept for at least 10 years in special cooling systems: these wastes/residues are too hot to be treated and disposed of in permanent storage areas. During this delicate phase dozens of cases of loss of radioactive material in the environment have been documented.

Every 1000 MW nuclear power plant produces about 30 tonnes per year of highly radioactive exhausted fuel. This is a complex mixture of solid, liquid, and gaseous discharges containing dozens of different isotopes. About 94% is uranium (almost all ^{238}U), 5% consists of various fission products (such as ^{137}Cs and ^{90}Sr), while the remaining 1% is composed of isotopes of plutonium or other artificial elements, likewise hazardous, such as americium-243 (^{243}Am). From a physicochemical viewpoint, the mixture is so complex that even if it were composed of non-radioactive material, it would still be very difficult to treat.

Since the late 1960s, the United States has sought to find a permanent and safe storage area for its radioactive wastes from its nuclear power plants and wastes originating from the dismantling of its nuclear warheads.

At the beginning this seemed a feasible task. After all, the United States is the most technologically advanced nation, the richest and most powerful on the planet, with vast, uninhabited, and geologically secure remote areas on its territory. The site initially identified was in Kansas, but then it was realized that the land in that area had been drilled countless times in search of gas deposits. The cover, so to speak, had been punctured, and so it was necessary to look for alternative sites.

Studies were started in 1978 on another site – the Yucca Mountain – a sort of natural bunker in the Nevada desert about 90 miles from Las Vegas. Millions of pages of reports and entire series of books have been written on the troubled history of this site. Initially, the site was to provide security guaranteed for 100 000 years; later it was reduced to *only* 10 000 years. A question that comes immediately to mind is: what is the sense of certifying something for a time period equal to twice the history of human civilization? The project was abandoned in 2009.

Nonetheless, the problem remained of how to transport thousands of tonnes of nuclear waste safely from its origin to the eventual storage area: by motorways or railways? On this aspect, also, no final decision has ever been reached. The estimated cost for the Yucca Mountain project amounted to 96 billion US dollars. Pending the construction of permanent storage sites, which will certainly take decades, there already exist approximately 70 000 tonnes of spent nuclear fuel in the United States that is constantly increasing and waiting to be disposed of safely.

At the current rates of production of electricity and nuclear weapons, the world would need a storage site with the capacity of a Yucca Mountain every two years. The United States has not been able to put one into operation in 40 years. During this period there have been various judicial investigations against public officials and private companies for corruption and for the falsification of documents.

Disposing of nuclear wastes and residues from civilian nuclear power plants safely and the management of huge sites contaminated by hundreds of nuclear tests carried out during the cold war (e.g., in New Mexico, in the Asian steppes, and in atolls of the Pacific) are among the more obscure and disturbing pages of the Earth's history of the last 50 years, as often stated by the environmental agency of the United Nations.

The use of the seas as nuclear waste disposal sites has gone on for decades. We fear the use is still practiced, although officially banned internationally. The trafficking of nuclear material, encouraged by the power vacuum that for years was featured in the ex-Soviet republics, is today very attractive to international criminal organizations.

The problem of securing highly radioactive wastes is undoubtedly one of the main obstacles to the expansion of the civilian nuclear industry. In this regard, more than thirty years ago Swedish physicist and Nobel Prize winner for physics, Hannes Alfvén, emphasized that:

> *The problem is where to dispose of the radioactive wastes that decay in hundreds or thousands of years. The geological sites must be absolutely safe because the toxic potential is tremendous. It is very difficult to meet this requirement for the simple reason that we have no experience with such long term projects. In addition, a permanent surveillance of these wastes requires a social stability for an unthinkable long time.*

Waste management of highly radioactive wastes has become an inter-generational problem.

We'll Settle the Bill Later

Uranium is a metal not found in any great abundance in nature, but is less rare than several metals of widespread use such as, for example, silver. The Earth's crust contains a mean concentration of about 3 parts per million (ppm) of uranium, so that each tonne of rock contains on average 3 grams. However, a cost-effective extraction of uranium necessitates finding minerals with much higher concentrations of uranium, in the order of hundreds of ppm (that is, about half a kg per tonne – slightly more than a pound per tonne).

The debate on exploitable reserves of uranium is controversial and contradictory. Numbers ranging from a few dozen to several hundred million tonnes are often quoted. The only certainty is that it is a finite resource, which will reach peak production and then decline. In theory, it is technically possible to extract uranium from the sea, where its concentration is about 3.3 parts per billion, or about 3 mg for each ton of water. However, the idea of recovering a fraction of the 4.5 billion tons of uranium dissolved in the oceans in an energetically and economically sound fashion is nothing but a wishful dream.

As shown in Figure 13, today's production of uranium is below demand since the nuclear crisis of the last thirty years has curbed investments in mining activities. As a result, the price of uranium (U_3O_8) rose more than seven times between 2002 and 2007; then it decreased about four times until 2010 following the economic crisis. Once again, later it increased by 50% around March 2011, then decreased by 20% after the Fukushima accident, reaching a steady value for several months (about 50 \$/pound, July 2012). The trend shows that the price of this resource is no less volatile than that of fossil fuels.

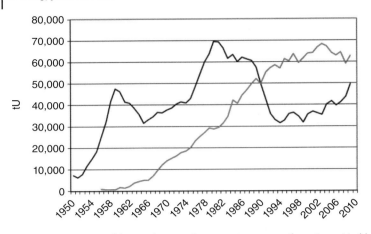

Figure 13 Evolution of the production of uranium in tonnes of uranium, tU, (black line) since 1950, and uranium demand (grey line) for civilian nuclear reactors. Until ~1970, production was almost exclusively for military purposes. Data source: World Energy Council 2010.

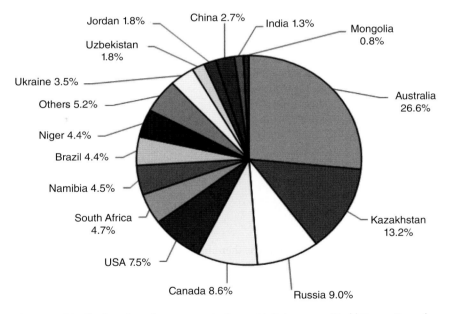

Figure 14 Distribution of uranium reserves in the world. Data source: World Energy Council 2010.

Among the top 15 owners of uranium reserves, there is not a single European Union country (Figure 14). Nevertheless, some people claim and insist that nuclear power is the route to Europe's self-sufficiency in energy, and to Italy's self-sufficiency. In this regard, it's worth recalling that the distribution of total energy consumption in Europe is: 23% from electricity and 77% from combustible fuels.

Nuclear power plants produce only electricity. Even if we produced all the electricity by the nuclear route, we would cover but a quarter of Europe's final energy consumption.

Often, we hear people talk about nuclear energy as though it consists of magical installations, creating energy from nothing, without emitting greenhouse gases and without producing any kind of pollution. The reality is very different, however. The industrial cycle of nuclear energy requires, in addition to a huge economic investment, large energy investments from fossil fuels, as also occurs for all alternative sources of energy including renewable ones.

The production of nuclear fuel, for example, is a long, complex, polluting, and energy-wasteful process. Major uranium mines in the world are located in remote areas. Extraction of minerals requires a lot of work and the use of huge excavators.

To extract the 160 tonnes of uranium needed to operate a standard nuclear installation for one year will require 160 000 tons of uranium-rich granite (1000 ppm in uranium) rock from the mines. Mines and quarries must be kept clear of water penetration, which is often drained into surrounding basins, with release of heavy metals and radioactive isotopes. Radioactive radon gas is ubiquitously present in underground mines.

To get *refined yellowcake*, which contains 80% uranium oxide – mostly U_3O_8 – the ore must be brought to an industrial plant to be crushed and treated with strong acids and other substances. The 159 840 tonnes of waste materials that remain and the huge quantities of chemicals used to treat them must be disposed of carefully because they contain radioactive isotopes.

To be used in reactors, uranium must first be enriched into the fissionable isotope ^{235}U, bringing it from its natural occurrence (0.7%) to about 3–4%. The oxide U_3O_8 is subsequently converted to uranium hexafluoride (UF_6), which is then subjected to energy-costly *ultracentrifugation* processes. The news of recent years, and the furore aroused by Iran's attempt to equip itself with these ultracentrifuges, explain how this is a key stage of the entire nuclear chain process. Whoever owns this technology – and there are very few such installations in the world – has the key to the final product: nuclear fuel.

The UF_6 is finally reprocessed by very complex chemical processes into UO_2 *bobby pins* as big as a cigarette filter, then inserted into 3.5-meter long zirconium bars (slightly less than 12 feet long) and a little more than a centimeter thick (a little more than 0.4 inches). A 1000 MW nuclear power station contains hundreds of bars of this type – they must be replaced and disposed of every three years.

There are high energy costs also downstream of the energy-producing chain. It is estimated that the process of decommissioning a nuclear power plant requires about 250 PJ, ten times more energy than it takes to demolish a gas-fed power station of equal power.

Payback time of an energy facility is the time required for the system to give back the cost of energy spent for its construction. For nuclear installations, this parameter is rarely taken into account. Researchers in Australia have recently estimated that with the current technology the *payback time* of a nuclear installation in

Australia would be about 7 years – that is, seven years operating at full capacity before the installation would pay back the energy spent to make it operational. The use of the conditional is a must because, even though Australia has significant reserves of uranium, it has never built a single nuclear power plant on its vast territory.

Current Nuclear Power Plants

The first commercial nuclear power plant in the world, Calder Hall in the North of England, was connected to the electrical network and began to produce electricity on October 17, 1956. Less than a year later, on October 8, 1957, there was a fire in an adjacent facility where plutonium was being produced. It was a very serious accident. The radioactive cloud spread over all the skies of Europe. The cooling towers of the Calder Hall station, one of the authentic icons of the nuclear age, were demolished on September 29, 2007.

During the course of more than fifty years, the Calder Hall site underwent significant expansion, becoming a large center for the reprocessing of spent nuclear material. It is now known as the Sellafield plant, an industrial complex responsible for the emission of unknown and unspecified quantities of radioactive isotopes, and has been the object of long-standing environmental and legal disputes.

As of the end of 2011, nuclear energy represented about 13.5% of the world's electrical energy and less than 6% of total primary energy. The construction of nuclear fission power stations experienced its golden age in the years spanning 1956 to 1986. In the last twenty years, the number of reactors in the world has remained essentially stable at around 435 units (Figure 15). The longevity of nuclear installations is therefore not insignificant – the average age is around 27 years (Figure 16).

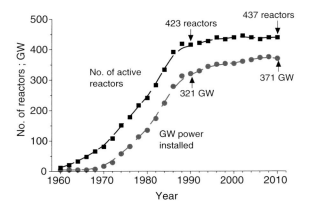

Figure 15 Time evolution of the number of the world's active nuclear reactors and global sources of electric power. The data are up-to-date to April 2011. After the Fukushima disaster, the number of operative reactors has dropped considerably (see Chapter 9).

Number of Operating Reactors by Age

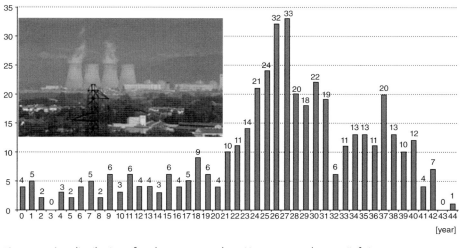

Figure 16 Age distribution of nuclear installations in the world as of September 15, 2011 (IAEA 2011). The age is calculated from the time a reactor was first connected to the electrical power grid. For the number of reactors, see the considerations made in Figure 15. Source: European Nuclear Society, http://www.euronuclear.org/info/encyclopedia/n/nuclear-power-plant-worldwide.htm; Photo by Muriel Boselli–http://www.reuters.com/article/2011/11/17/us-france-nuclear-tests-idUSTRE7AG0HQ20111117.

There are currently about 60 nuclear reactors under construction worldwide that will not replace those that will be decommissioned for reasons of age. In recent years, the number of new reactors has not exceeded the number of decommissioned reactors. These figures are in sharp contrast to erroneous claims that the boom of nuclear energy is ongoing. The landscape of operating nuclear reactors has been particularly stagnant in regions where the largest number of installations are located, namely in the United States and in the European Union. There are only two new reactors under construction in Western Europe, both of which on sites with existing old nuclear power plants.

Despite this freeze in the number of installations, the quantity of available power from nuclear stations has grown over time, due mainly to a more efficient management of existing facilities. In 1973, nuclear installations operated 50% of the time – the other 50% was devoted to maintenance. Today these installations operate at more than 80% of the time despite their significant age.

Nuclear technology has evolved considerably since 1956. The vast majority of the approximately 430 reactors now in operation belong to the *second generation* of various types. Very popular are the *light water reactors* (LWR), which use ordinary water to slow down the neutrons and to cool both the installation and the enriched uranium fuel rods. It was with this principle in mind that the Three Mile Island reactor was built, whereas the Chernobyl nuclear complex used water for the cooling process, and employed graphite as the moderator of neutrons. The Chernobyl-type nuclear power plants are subject to significant wear and corrosion, not to mention their costly and delicate maintenance.

There are also several nuclear installations that use *heavy water* (water containing deuterium, a heavier isotope of hydrogen). The most common is the Canadian CANDU reactor technology. These reactors use natural uranium (i.e., not enriched); however, compared to the LWR reactors they produce greater quantities of spent nuclear fuel that is particularly rich in plutonium.

Currently, the more modern installations are defined as *third generation*. They consist basically of LWR nuclear reactors based on a more evolved technology and, most importantly, on a more competitive economic base.

Tomorrow's Nuclear Power Plants (Maybe)

Nuclear power plants of the future should belong to the so-called *fourth generation* or G4. Their development is currently the subject of a protocol of international collaboration among thirteen countries. It's relevant to emphasize that G4 reactors will likely be built not before 30 years.

There are no less than six different types of G4 projects being examined actively today, linked by at least three common objectives: (i) increase the conversion yield of the nuclear-to-electricity installations, which currently is around 30%; (ii) build safer nuclear power plants less exposed to possible indirect military uses (*nuclear proliferation*); and (iii) make the nuclear energy source economically competitive with traditional and/or renewable energy sources.

Reconciliation of these three requirements is going to be a difficult challenge, and the success of the undertaking is by no means guaranteed. To increase the yields, for example, the nuclear plants will have to operate at temperatures between 500 and 1000 °C (today they operate at around 300 °C). This will require materials highly resistant to high temperatures and to radiations, and therefore will no doubt lead to increased costs.

Four of the G4 projects involve the reprocessing of spent fuel, which is economically wasteful and produces plutonium. Reprocessing the spent fuel should be carried out so as to make it *resistant to proliferation*, or otherwise difficult to steal and process. Absolute security of these activities can never be fully guaranteed, however, since a minimal technical and/or political indiscretion cannot be fully avoided.

One of the factors critical to the long-term prospects of nuclear energy is the availability of uranium, a resource limited to no more than 50 years with current technologies and consumption levels. G4 projects aim at the use of ^{238}U – even if obtained from spent fuel – that can be converted into plutonium by bombarding it with fast high-energy neutrons.

In some configurations, G4 projects will be of the *breeder-type* (so-called *breeder reactors*) that produces its own fuel. This is possible when the reactor operates with plutonium and added natural uranium that, as we have seen, can be transformed into weapons-grade plutonium as a consequence of the ensuing fission reactions. Attempts to develop safe, reliable and economically viable breeder reactors have been ongoing for decades. The outcome has so far been a failure, starting with the famous French Super-Phoenix installation in which Italy has participated actively through its electric utility ENEL. The installation was closed permanently in 1997 after 12 years of troubled operation within the framework of a colossal economic fiasco.

The eventual success of fourth generation projects could virtually provide an unlimited availability of future fissionable material. But this will not resolve the problem of fuel handling and the safe disposal of the radioactive wastes.

The Harsh Reality of the Marketplace

Nuclear energy has a certain charm for its power and its technological elegance. Nonetheless, it still has its limitations and its unknown consequences: (i) the economy of the industrial cycle, (ii) the safety of the power plants under ordinary conditions and in the presence of catastrophic scenarios (earthquakes, terrorist attacks, etc . . .), (iii) waste disposal, and (iv) its indisputable and ambiguous link to the military establishment. No other industrial activities would be allowed to continue their growth without having resolved these serious shortcomings.

Faced with a picture so complex, it is frightening to listen to debates of incredible superficiality that claim that the nuclear option is a clean and cost-effective solution for a country's energy demand. Those in Italy that support this thesis put forward the argument that Italy is constrained to buy electricity at a high price from existing nuclear power plants in France. This is far from the truth. In fact, it is France that is forced to sell its electricity supplies to neighboring countries at low cost during the night periods so as to get rid of its surplus. Nuclear power plants cannot simply be turned on and off by a simple turn of a switch in order to keep in step with the discontinuous and evolving daily demand for electricity.

In a climate of confrontation between those in favor of the nuclear option and those that oppose it, there are many in Italy who believe that the nuclear option is off the table because of the strong opposition by environmentalists. Analysis of the history of the last fifty years tells a very different reality. Certainly, the major accidents of the Fukushima, Chernobyl, and Three Mile Island nuclear installations have undermined the social acceptability of the nuclear technology, concerning which even Fermi had shown some serious reservations. The crisis of the nuclear option is primarily an economic issue, not a technological one.

Liberalization of the electricity markets has been a formidable deterrent to investments in the nuclear option, however, demonstrating that nuclear energy would not survive on the free market. Unless generous state coffers can guarantee the enormous costs of the industrial cycle, in particular those upstream and downstream – that is, construction and decommissioning of power stations – no private investor will be stupid enough to invest in projects that face a number of risks: for example, long and costly legal battles with local communities at the selected sites.

Despite generous government support, the rumored renewed launch of nuclear programs in the USA has not met with excessive enthusiasm. Yet the Obama Administration recently gave the OK to build two nuclear power plants. Another problem is related to construction times that are no less than 10 years, the traditional period for the most efficient countries: it often takes 10 years just to build a school in Italy! This implies an enormous financial burden if such an undertaking were to be realized.

The incident that occurred in 2011 in Japan, a country that many believed technologically advanced and foolproof, has proven once and for all that nuclear technology is too complex to be totally *predictable*. As some analysts had predicted, the Fukushima disaster marks the final sunset of the revival of nuclear energy, especially in countries where the market economy dominates. In short, the civilian nuclear industry, moribund for the last thirty years, has been abandoned not only by market pressures but also by entrepreneurs, and this even before the environmentalists had anything to say on the matter.

Solution or Problem?

In light of the enormous human and material damage that climate change causes or could cause, some scientists continue to propose nuclear energy as a solution to resolve the energy issue. They view the nuclear option as the energy source that offers reduced intensity of greenhouse gas emissions relative to those from fossil fuel power stations.

If we hypothesize that nuclear power can play an important role in the global energy system between now and 2050, however, as a minimum the scenario should be the following: (i) replacement of all current reactors in operation (about 430) that have reached their age limits, (ii) substitution of 50% of current coal-fired power plants, and (iii) coverage of 50% of the new demand for electricity. This scenario would involve the construction of about 2500 nuclear installations with a capacity of 1000 MW each, that is, one would have to be built every week between now (2012) and 2050. This is completely unrealistic. Technically speaking, the times are too short, new sites of possible uranium deposits are unknown, and adequate sites for the disposal of nuclear wastes are lacking, not to mention the thousands of tonnes of plutonium that would be produced.

For the rest, all authoritative analyses predict a limited role for nuclear energy in any future energy scenario. In this regard, the International Energy Agency (IEA) predicts that in 2030 the nuclear option will provide a share of world electricity requirements substantially less than current levels. This figure represents no less than a defeat for a technology that in recent decades has burned over 60% of the research and development funds in developed countries that could have been better devoted to new energy technologies.

With the impending threat of climate and energy crises, then, it seems unreasonable to continue to spend huge amounts of intellectual energy and financial resources to follow the slowest, the most expensive, the riskiest, limited, and rigid path taken so far. Is it not time to get out of the 1942 box and undertake – with greater gusto – research on renewable energy technologies?

Nuclear Fusion: if Not Roses . . . Then What?

Unlike nuclear fission, nuclear fusion is the process that fuels the Sun and other stars, wherein two light atomic nuclei fuse together to form a heavier nucleus, with a mass less than the sum of the masses of the starting nuclei. Energy is liberated in the form of electromagnetic radiation and kinetic energy of the products according to Einstein's law (Figure 17). The best candidates to reproduce this process on Earth are deuterium and tritium (two isotopes of hydrogen). One gram of these hydrogen isotopes could produce an amount of energy through the fusion process equivalent to that released by 11 tons of coal.

Nuclear fusion has been exploited, albeit in an uncontrolled form, in the hydrogen bomb – the H bomb. To produce energy it will be necessary to construct thermonuclear reactors in which the fusion process takes place in a controlled manner. As we speak, research in this field is still at the preliminary stage.

In the long run, nuclear fusion could be the radical solution to the energy problem. In theory, it could produce limitless amounts of energy with reduced harmful gas emissions or greenhouse gases and would generate radioactive wastes with half-lives of the order of tens of years limited mostly to reactor components.

It was predicted in 1972 that in the year 2000 electricity would be generated by commercial nuclear power stations based on nuclear fusion. This forecast has

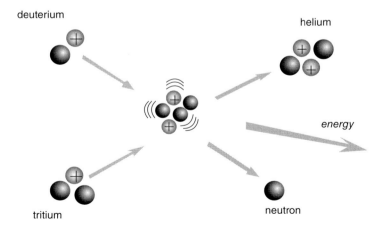

Figure 17 An outline of the process of nuclear fusion: a nucleus of deuterium and tritium are combined to form a helium nucleus, releasing energy.

proven far too optimistic. Experiments conducted thus far have produced at most a few dozen MW of power for short periods. Currently, the process consumes more energy than it produces.

The most promising technique for exploiting nuclear fusion uses the infrastructure referred to as *Tokamak*, with large donut-shaped vacuum chambers. A mixture of deuterium and tritium is injected and heated to produce a plasma (i.e., an ionized gas), which is then confined by intense magnetic fields to prevent it from cooling when coming in contact with the reactor walls.

The *International Thermonuclear Experimental Reactor* (ITER) project is based on this approach. The project started in the South of France with the participation of the major energy-consuming countries. Initially, the project was to build a first demonstration of a 500-MW power plant by 2030, followed by commercial installations by 2050. As is always the case for nuclear megaprojects, however, costs have soared year after year and estimated completion times have got longer and longer. Both the United States and the European Union have already begun to reduce funding for such an enterprise.

Many scientists remain skeptical about the feasibility of nuclear fusion. In any case, even if one day the nuclear fusion cavalry came to the rescue, it will occur well beyond the time threshold that can guarantee a painless way out from the era of fossil fuels.

7
Energy from the Sun

Scientists are called to see what everyone has already seen and to think what nobody has yet thought of.

Albert Szent-Györgyi

Spaceship Earth is not an isolated system. It must meet many of the needs of its crew with whatever resources (non-renewable) it has in its hold. However, Earth also enjoys a great benefit as it orbits around the Sun. It continuously receives an immense quantity of energy in the form of electromagnetic radiation – light and heat.

It is this energy that renews some of the fundamental resources of the Earth, such as the products of the plant world (including oxygen) through the process of photosynthesis, and with them all the food pyramids: drinking water through evaporation and subsequent precipitation (rain), and the wind through the formation of temperature gradients that cause the movement of large air masses.

Fossil fuels are also a resource continually *renewed* by sunlight, since they derive from the process of photosynthesis. In practice, for us, however, they are non-renewable resources because their formation requires a (geological) period a million times longer than the time needed to consume them.

Solar energy is abundant. In less than an hour, the Earth receives from the Sun an amount of energy equal to the entire world's annual energy consumption. Solar energy, unlike fossil fuels, is present in all regions of the planet, although with significant differences depending on the latitude. And because the Sun will shine for a few billion years more, besides being abundant and well distributed, solar energy is also an inexhaustible energy source in our time scale.

These very important qualities are, however, mitigated by two defects: the intensity of solar energy Earth receives is low and is intermittent on a local scale, because it depends on weather conditions and on the alternating day and night cycles.

The current lifestyle in developed countries requires a power density ranging from 20 to $100 \, W/m^2$ for a home to $300–900 \, W/m^2$ for a steel mill.

The power density of solar energy is on average approximately $170 \, W/m^2$, a value that decreases drastically when converted into usable power. Therefore, it could never operate steel mills and other facilities of high energy consumption – for example, hospitals – with the energy of sunlight that falls on a day on their rooftops.

Powering Planet Earth: Energy Solutions for the Future, First Edition. Nicola Armaroli, Vincenzo Balzani, and Nick Serpone.
© 2013 Wiley-VCH Verlag GmbH & Co. KGaA. Published 2013 by Wiley-VCH Verlag GmbH & Co. KGaA.

It is easy to guess, then, that the main challenge of science and technology is to store somehow this gigantic but nonetheless dilute solar energy flux for later use at intensities needed where and when required.

Conversion and Exploitation of Sunlight

As noted earlier, energy is much more useful when it is concentrated (but not too concentrated, otherwise it becomes hazardous), storable, and transportable. The reasons are simple:

1) Energy in concentrated form is required to meet the needs of large and complex structures.

2) If it is storable, energy can be accumulated and ready for use.

3) If it is transportable, energy can be used in places other than those in which its reserves are located. In the case of transport over long distances, in particular by air or by sea, it is absolutely necessary that the energy be in a form that can be transported.

Combustible substances, such as fossil fuels, meet all three requirements, albeit to varying degrees. Electricity meets the first and the third—the second would be possible with more efficient accumulators. Thermal energy meets at most the first requirement, when it is in the form of high-temperature heat.

The Sun's energy can be converted into low-temperature heat, and although not impossible, it is nonetheless difficult to convert it into high-temperature heat, electricity, and even more so into fuels.

From Light to Heat

Conversion of solar energy to heat at low temperature can be achieved using *solar panels*, which are not to be confused with photovoltaic panels used for the generation of electricity—the latter will be discussed later. These panels serve as heat collectors wherein a liquid flowing in copper pipes is heated by sunlight and is then used to exchange heat with a water reservoir (see Figure 18).

At Italy's latitudes, a panel of about $3\,m^2$ is sufficient to provide domestic hot water for an average family of four. With more extensive surfaces, it is possible to provide heat to a heating system of larger buildings.

The lifetime of a solar panel is at least 30 years, requires only little maintenance, and within two years produces an amount of energy equal to that which was required to manufacture it. Such panels are simple, reliable and inexpensive thanks to incentives promoted by many Governments. In some European regions, local laws require new detached houses to have solar (non-photovoltaic) panels installed.

At the end of 2011, the world's power from solar panel installations amounted to 232 GW, a 20% increase from the previous year. In 2010, the area of solar col-

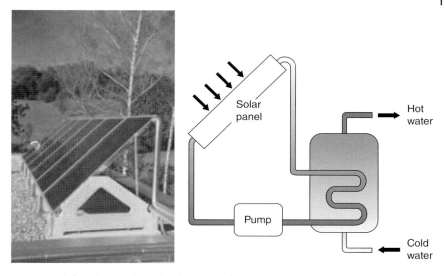

Figure 18 (left) Solar panels used to heat water for domestic use; (right) outline of their operation. Photos: Fritz/Shutterstock.

lectors installed in China alone reached the $150\,km^2$ mark (an area nearly equal to that of the city of Milan) – it is expected to double by 2020.

The use of solar panels will also, indirectly, save electricity. Note that modern houses consume large amounts of electricity to heat the water in washing machines, dishwashers, showers and the like.

The use of solar panels for heating domestic water in Italy is still desperately low. In 2011, the surface equipped with solar panels per inhabitant was 15 times less than that of *colder* Austria and 11 times smaller than that of Greece. It should be noted that most of the Italian solar installations are located in the South Tyrol region (Italian Alps), and not in the sunny South. In fact, even the hot water used in showers in Italian resorts is almost always obtained by combustion of liquid propane gas (LPG). This is a symptom of a country absolutely unable or politically unwilling to look to the future.

From Light to Electricity

Direct conversion of light to electricity takes place in *photovoltaic cells* (PV cells) wherein a semiconductor material (usually silicon) absorbs the sunlight and gives rise to a movement of electrical charges. Individual photovoltaic cells connected together form modules, which when assembled into larger devices, are called *photovoltaic panels* (Figure 19). This technology has been in use for some time in remote areas far from electric transmission networks (such as high mountain huts and artificial satellites) or in gadgets that require small amounts of energy (e.g., small calculators and watches). To meet the needs of a single

Figure 19 Solar panels that convert sunlight into electricity. Photo: Tom Grundy/Shutterstock.

average family at the latitude of Italy would require an area of about $18 \, m^2$ of photovoltaic panels.

At current levels, the photovoltaic industry is growing at a rate of about 70% per year, but still produces a relatively small amount of energy when compared with that produced from fossil fuel power plants or from nuclear power stations. About 68 GW of power was produced by the end of 2011 from worldwide power installations.

The duration of current photovoltaic silicon panels is about 30 years, and return the energy needed to build them is approximately 1 to 3 years. The cost of electricity produced by photovoltaic panels is still greater than the cost of electricity produced with fossil fuels. Cost parity is expected to be reached in a few years. Prices are bound to decrease as production increases. It's worth pointing out, though, that photovoltaics (PVs) would already be competitive economically if externalities associated with fossil fuels described earlier were also taken into account.

Interestingly, some African countries are experiencing high levels of growth based on photovoltaics. This technology allows energy to be produced where needed, without an expensive long distance distribution network. It is therefore particularly suitable for poor countries, that is, poor in economic resources necessary to build large electrical infrastructures. It's worth remembering that there are still 1.5 billion people worldwide who have no access to electrical power.

The world's PV industry leaders are China, Taiwan, Germany, and Japan, which today harvest the fruits of far-sighted industrial innovation that began more than 20 years ago. The United States is trying to recover lost ground, making a quantum leap into a second generation of photovoltaic-based thin-film solar cells.

Currently, some industries produce this type of cells using a technique similar to that with which newspapers are printed. Placed on aluminum foil is a type of ink that contains a mixture of semiconductor nanoparticles made of copper (Cu), indium (In), gallium (Ga), and selenium (Se). Further research should lead to solar cells deposited on sheets of folding plastic or even embedded in paints. Many predict that the photovoltaic technology will be the next *disruptive technology* that will change dramatically the way in which electricity is produced and distributed.

Photovoltaic systems are usually placed on rooftops of houses and factories – they can be either isolated or connected to the distribution network. In the first case, systems employ batteries to accumulate the energy produced during the day for nocturnal uses. In the second case, the systems exchange energy with the network, selling the day's surplus and buying requirements for the night hours. At the end of the year, the difference is collected, with a special fee for the share produced – that is, the consumer invoices the company/producer of electricity and not *vice versa*. In so doing, the consumer can recover the investments of the PV installation in less than 10 years.

The production of photovoltaic electrical energy needs wide open spaces, but these need not be as extensive as one might think. It has been calculated, for example, that using panels with a 10% conversion efficiency (already exceeded) to cover an area of about $26\,000\,km^2$ (or about $10\,000$ square miles) of photovoltaic modules would suffice to produce all the electricity consumed by the United States. This is certainly an extensive area, but it's still less than a quarter of the area covered by the interstate highway network. A recent study indicates that to cater to the European need for electricity with photovoltaic panels would require a space of about 0.6% of the area of the various countries (Figure 20). For instance, Italy would need $2400\,km^2$ (i.e., 927 square miles) – an area the size of a small Italian province.

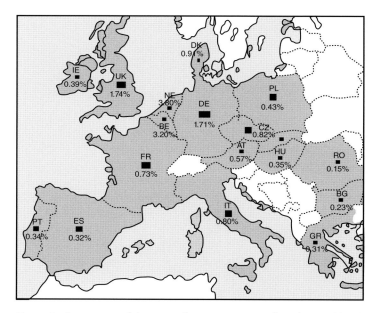

Figure 20 Percentages of the areas of European nations that, if covered by photovoltaic panels, would be sufficient to provide all the electricity required by the given nation. Data source: *Solar Energy* **81** (2007) 1295–1305.

To produce *all* the energy (electrical and non-electrical) consumed in the United States by photovoltaic devices would need the use of 2.7% of its total land mass. This percentage increases to 24% for Belgium and drops to 0.3% for Brazil.

Concentrating Sunlight

The conversion of solar energy into electricity can also be achieved by exploiting the mechanical energy of steam generated by a liquid brought to a high temperature (via turbines and alternators, as in conventional power stations). Such high temperatures can be achieved by focusing the sunlight onto a boiler by a system of mirrors or, alternatively, by using a parabolic linear collector system that concentrates light onto a tube into which flows a liquid that absorbs the Sun's heat. The first method was re-activated recently, while the second (more efficient) method is under development. Various facilities are under construction in Spain and in the United States. There is also one at Priolo, Sicily, the result of a joint venture between ENEL (Italy's largest electrical utility) and ENEA (Italy's National Agency for Renewable Energies).

Another way in which high-temperature heat from the sun could be used is to produce hydrogen from the thermal splitting of water. The possible use of hydrogen in the energy field will be discussed shortly.

Light to Chemical Energy – *Natural Photosynthesis*

Natural photosynthesis is a process that takes place in plants: sunlight is absorbed by chlorophyll molecules. Through their mediation, the process transforms substances of low energy content – H_2O (water), CO_2 (carbon dioxide) – into high energy substances – O_2 (oxygen), and carbohydrates contained in vegetable products (reaction 1).

$$H_2O + CO_2 \xrightarrow[\text{Chlorophyll}]{\text{Solar radiation}} O_2 + \text{carbohydrates} \tag{1}$$

Natural photosynthesis, which ultimately turns sunlight into chemical energy, is also the process that indirectly has given us deposits of fossil fuels (coal, oil and natural gas). These substances were formed underground following the transformation of plant and animal organisms through a series of complex chemical processes that occurred in the course of hundreds of millions of years. Natural photosynthesis continues to produce fossil fuels, albeit at a pace immensely slower than the pace at which these fuels are consumed.

Natural photosynthesis can convert up to about 5% of the energy of sunlight into chemical energy, but on average only 0.3% of solar energy that reaches the Earth's surface is converted into vegetation, of which only a small fraction is harvested and exploited.

The natural photosynthetic process is a very complex one; its mechanism was revealed thanks largely to studies carried out in the last few decades. Simply stated, the early stages of the process are:

1) Light is absorbed by the leaves through an organized system of chlorophyll molecules yielding molecules in their excited electronic state.

2) The electronic energy so collected, a bit like an antenna, is then transferred onto a specific site – the *reaction center*.

3) On this site, the energy is used in extremely short times – of the order of picoseconds, 10^{-12} seconds – to separate charges of opposite sign: *plus* on one side and *minus* on the other.

Subsequent to this first event, the process in plants continues through a very complex series of reactions that lead to the formation of molecular oxygen and carbohydrates. Everything happens thanks to a precise molecular organization, the result of billions of years of evolution: (i) organization in *space* – correct distances between the various molecules involved in the process, (ii) organization in *time* – some reactions are faster than others and take place in extremely short times, and (iii) organization in *energy* – each stage of the process uses a part of the energy provided by sunlight.

Light to Chemical Energy – *the Sunshine Vitamin*

Vitamin D – also known as *calciferol* – is the general name for a collection of steroid-like substances including vitamin D2 (*ergocalciferol*) and vitamin D3 (*cholecalciferol*). Vitamin D is produced endogenously when ultraviolet rays from sunlight are absorbed by the skin and trigger its synthesis. It is a fat-soluble vitamin naturally present in a very few foods, otherwise often added to foods – it is also available as a dietary supplement.

Vitamin D is biologically inert, and so it must be transformed into its active form by two hydroxylations in the body – what chemists would refer to as adding two OH groups to the inert Vitamin D molecule. The first transformation takes place in the liver and converts vitamin D into a product called 25-hydroxyvitamin D – also known as 25(OH)D or as *calcidiol* (see below). The second transformation takes place primarily in the kidneys and forms the physiologically active product 1,25-dihydroxyvitamin D – *calcitriol*, also known as 1,25(OH)$_2$D.

Deficiency of vitamin D can not only cause rickets among children but also precipitates and exacerbates osteoporosis among adults and causes the painful bone disease known as osteomalacia. Vitamin D deficiency has also been associated with increased risks of deadly cancers, cardiovascular disease, multiple sclerosis, rheumatoid arthritis, and type 1 diabetes mellitus.

Maintaining blood concentrations of 25-hydroxyvitamin D above approximately 30 ng/mL is important for maximizing intestinal calcium absorption and also for providing the extra renal 1-alpha-hydroxylase present in most tissues to produce the compound 1,25-dihydroxyvitamin D3. Although chronic excessive exposure to sunlight increases the risk of non-melanoma skin cancer, the avoidance of all direct sun exposure can lead to an increase in vitamin D deficiency, which can only lead to serious consequences.

The amount of vitamin D produced depends on the intensity of the UVB radiation from the Sun and on many other factors: season, time of day, length of day, cloud cover, smog, skin melanin content, and sunscreen are among the factors that could affect UV radiation exposure and therefore vitamin D synthesis (Scheme 1).

Darker-skinned individuals may need 5 to 10 times more sunlight exposure than a fair-skinned person to make the same amount of vitamin D, typically between 3000 and 20 000 International Units (1 IU = 40 micrograms). In northern climates, sunlight is too weak in parts of the year for the body to make any vitamin D – a period referred to as the *Vitamin D Winter*.

Surprisingly, however, a geographic latitude does not consistently predict average 25(OH)D levels in a population. Nonetheless, opportunities exist to form vitamin D – stored in the liver and fat – from exposure to sunlight during the spring, summer, and fall periods even in far northern latitudes.

Complete cloud cover can reduce UV energy by 50%, whereas shade reduces it by 60%. UVB radiation does not penetrate glass, so exposure to sunshine indoors through a window will not produce the necessary vitamin D in your body. Sunscreens with a sun protection factor (SPF) of 8 or more can block vitamin D-producing UV rays. However, to the extent that most people generally do not apply sufficient amounts of sunscreens to cover all sun-exposed skin and achieve the recommended SPF factor, vitamin D synthesis can still occur even when skin is (partially) protected by sunscreen.

It has been suggested by some vitamin D researchers that about 5 to 30 minutes of exposure to sunlight between 10 AM and 3 PM – at least twice a week – to the face, arms, legs, or back without sunscreen is usually sufficient for vitamin D synthesis. We hasten to note that despite the importance of sunlight for vitamin D synthesis, people must limit exposure of skin to sunlight for long periods as UV radiation is a well-known carcinogen responsible for most of the estimated 1.5 million skin cancers and the 8000 deaths due to metastatic melanoma that occur annually in the United States.

The few foods that are thought to be best sources of vitamin D are the flesh of fatty fish – for example, salmon, tuna, and mackerel – and fish liver oils. Small amounts of vitamin D are also found in beef liver, cheese, and egg yolks. However, vitamin D in these foods is primarily available in the form of vitamin D_3 and its metabolite 25(OH)D3. Some mushrooms provide this vitamin as vitamin D2 in variable amounts.

Vitamin D promotes absorption of calcium in the gut and maintains adequate serum calcium and phosphate concentrations to enable normal mineralization of bone and to prevent hypocalcemic tetany – a disease caused by an abnormally low level of calcium in the blood. It is also needed for bone growth and bone remodeling by osteoblasts and osteoclasts. An osteoblast is a cell that makes bone by producing a matrix that then becomes mineralized – bone mass is maintained by a balance between the activities of osteoblasts that form bone and other cells called osteoclasts that break it down. Without sufficient vitamin D, bones can become thin, brittle, or misshapen. Vitamin D sufficiency prevents rickets in children and osteomalacia in adults. Together with calcium, vitamin D also helps to protect older adults from

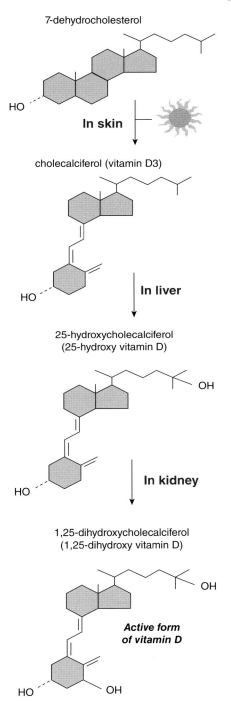

Scheme 1 Steps in the synthesis of Vitamin D. Source: http://www.vivo.colostate.edu/hbooks/pathphys/endocrine/otherendo/vitamind.html.

osteoporosis. Vitamin D also has other roles in the body such as modulation of cell growth, neuromuscular and immune function, and reduction of inflammation.

Biomass and Biofuels: Yes, but . . . !

Currently, the natural photosynthetic process produces annually approximately 230 billion tons of terrestrial and marine plant mass. This biomass is a real *solar fuel*, storable and transportable, but with an average energy density much less than that of fossil fuels.

Biomass can be burned to produce heat. At present, however, biomass arouses much interest in that it can be used to produce *biofuels* that can be used jointly with fossil fuels or used alone to replace fossil fuels. *Biodiesel* is produced primarily by chemical treatment of various kinds of vegetable oils–for instance, rapeseed and sunflower.

Bioethanol is obtained by fermentation of agricultural products rich in carbohydrates and sugars, such as corn, sugar cane, and sugar beet. Brazil uses sugar cane, whereas maize is used in the United States. Bioethanol can be used pure in specially modified engines, or can be used mixed with gasoline. Brazil has replaced nearly 40% of gasoline for its transportation network. However, it is doubtful that the same can be achieved in other parts of the planet, for the peculiar climatic and geographic conditions of Brazil are hardly found elsewhere.

Actually, the importance of biofuels is not easy to evaluate for a variety of reasons outlined in Figure 21:

- to produce biofuels it is necessary to undertake no less than ten different processes (prepare fertilizers, distribute them on the ground, plowing, harvesting, transport, and so on) that require a high consumption of energy, normally

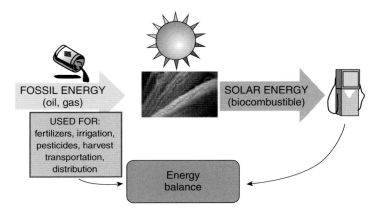

Figure 21 The production of biofuels requires, in addition to the use of fertile land, large quantities of fossil energy.

obtained from fossil fuels. Authoritative scholars argue that, on balance, the energy input to produce certain biofuels (e.g., ethanol from maize) exceeds that obtained from the use of the same biofuel;

- vast tracts of land and huge quantities of water are required for irrigation to obtain large quantities of biofuel;

- extensive cultivation for energy purposes can lead to the destruction of valuable ecosystems for the balance of the biosphere – for example, the rainforests;

- it is difficult to estimate what advantage or benefit biofuels[1] bring to reduce greenhouse gas emissions, considering the large amount of energy (mainly oil and gas) needed for the growth and processing of biomass;

- in a world where hundreds of millions of human beings still suffer from hunger, it is morally questionable to grow products to feed mechanical machines of other human beings who are already rich and satiated. For example, cost increases were recorded between March 2007 and March 2008 for corn (31%), rice (74%), soybean (87%) and wheat (130%) attributable, at least partly, to the increasing use of agricultural land for the production of biofuel precursors. It would be better to produce ethanol from cellulose (second-generation bioethanol) using certain types of plants as starting materials that grow in uncultivated land. However, this process is still in preliminary experimentation.

Each of the above topics would require a rather lengthy treatment, especially as the industry is in a tumultuous developmental state. Research studies are ongoing to obtain *biobutanol* from bacterial fermentation processes, whereas biodiesel is obtained from algae and even from animal fat.

The development of biofuels has certainly interesting prospects. However, it will probably make only a small contribution to solving the energy problem. It is estimated that if Europe and the United States wished to replace only 5% of their consumption of gasoline and diesel with biofuels produced with the technologies currently available, they would have to use 20% of their arable land.

The European Union has recently established that by 2020 biofuels will have to contribute up to 10% to the transportation sector. A document from the Royal Society (UK) maintains that this proposal is unworkable.

Artificial Photosynthesis

At the beginning of the last century, when crude oil and natural gas were not yet in common use, domestic and industrial growth was based essentially on the

---

1) In response to an article published in 2008 in the authoritative magazine *Science* a lively debate ensued on the sustainability of the production of biofuels. According to some scientists, the environmental and energy balance is strongly negative, yet the market is artificially kept alive by strong public incentives.

consumption of enormous quantities of coal, causing major problems of air pollution. Even then, some scientists wondered why mankind should make use of the *dirty solar-based fossil fuel* and not the clean and abundant energy that Earth receives continuously from the Sun to meet this growing energy need. Among these scientists, Giacomo Ciamician, then Professor at the University of Bologna, played an important role in the debate. Incidentally, the Chemistry Department at the University of Bologna is named in his honor.

At the 8th International Congress of Applied Chemistry, held in New York City in 1912, Ciamician presented a lecture on The Photochemistry of the Future. He faced the energy problem with some striking strong words which showed some of his foresight. He stated, among other things:

> *Modern civilization is the daughter of coal. Modern man uses it with increasing eagerness and thoughtless prodigality for the conquest of the world. The Earth still holds enormous quantities of it, but coal is not inexhaustible. The problem of the future begins to interest us.*

Fascinated by the ability of plants to make use of sunlight, Ciamician anticipated the day when their secret would be revealed and used by man to solve the energy problems:

> *Forests of glass tubes will extend over the plants and glass buildings will rise everywhere; inside of these will take place the photochemical processes that hitherto have been the guarded secret of the plants, but that will have been mastered by human industry which will know how to make them bear even more abundant fruit than nature, for nature is not in a hurry and mankind is. And if in a distant future the supply of coal becomes completely exhausted, civilization will not be checked by that, for life and civilization will continue as long as the sun shines!*

If the word "coal," which was the only fuel used at that time, was replaced by "fossil fuels," such a statement would hold even today. Both the nervousness and the restlessness of our civilization, which Ciamician noted, have increased in the face of problems that appear too complex to be managed – pollution of the biosphere, the greenhouse effect, the growing inequality in the distribution of wealth, increased population, and the widespread depletion of natural resources.

The secret of natural photosynthesis has now been revealed and understood. But mankind, which is in an even greater hurry, has not yet managed to use it to produce artificial fuels through the conversion of solar energy. The realization of Ciamician's dream is one of the most important challenges that science faces today to survive the looming energy and ecological crises.

Research on artificial photosynthesis is aimed at producing fuels from two widely distributed substances: water and carbon dioxide. For example, water would have to be split into hydrogen and oxygen, according to reaction 2:

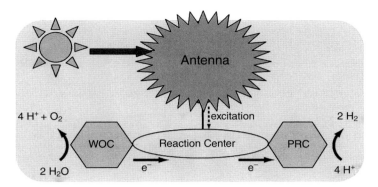

Figure 22 Cartoon illustrating the process of an artificial photosynthetic system for water splitting; WOC = water oxidation catalyst; PRC = proton reduction catalyst. Adapted from D. Gust, T.A. Moore and A.L. Moore, *Faraday Discuss.* 2012, 155, 9–26. Copyright 2012 by The Royal Society of Chemistry.

$$H_2O \xrightarrow[\text{Artificial molecules}]{\text{Sunlight}} H_2 + 1/2\, O_2 \qquad (2)$$

Carbon dioxide might instead be reduced to methanol (CH_3OH), with concomitant generation of molecular oxygen. But since the latter process is already very complicated, even on paper, all the attention of researchers is focused on splitting water into hydrogen and oxygen, which would create a closed loop in energy production. We start with water, a relatively inert molecule of low energy content (and therefore abundant on Earth), which when injected with energy in the form of sunlight leads to the separation of the two components that comprise it, namely hydrogen (a fuel) and oxygen (an oxidizing agent) – see Figure 22. When these two components are recombined in a combustion process (or in a fuel cell) they produce thermal energy (or power) by releasing the stored energy and forming water as the sole product. To split the water molecule, however, requires the intervention of substances that are capable of absorbing sunlight and mediating the process, as does chlorophyll in natural photosynthesis.

Some partial goals have been achieved recently, but there are many problems not yet resolved: for example, finding robust and efficient catalysts capable of intervening in processes involved in the generation of hydrogen and oxygen at the end of the sequence of reactions. It will take several years and the work of many scientists to achieve the production of hydrogen from water via the photochemical route. In the meantime, however, hydrogen has become a buzz word and a legend in the mass communication media.

The Hydrogen Myth

According to some newspaper reports, hydrogen will solve all the energy problems. It is a clean, abundant, and even *democratic* energy source. In poorly informed scientific circles, and in political circles that document themselves only through newspapers, we often hear speeches that sound more or less like this:

> *Today we are forced to use fossil fuels which are continuously being depleted, produce carbon dioxide, and are the cause of the greenhouse effect. In a few years, however, we shall finally use hydrogen, which does not pollute because when it's used it produces only water.*

Unfortunately, things are far more complicated. The so-called *hydrogen economy* is in fact a very complex problem, for which it is difficult to find a quick answer. Let's examine why this is so.

Hydrogen, when used to produce energy is molecular hydrogen, a gas with formula H_2. It's been known for more than two hundred years that when hydrogen burns it releases energy, just as happens when you burn natural gas, oil, or coal. The big difference, however, is that while the combustion of fossil fuels produces carbon dioxide, hydrogen produces only water (reaction 3):

$$H_2 + 1/2\, O_2 \rightarrow H_2O + \text{energy} \tag{3}$$

There is also another, often forgotten key difference between hydrogen and fossil fuels. Fossil fuels are *primary* energy sources, found in natural deposits from which they are extracted and then used. Earth, however, has no deposits of molecular hydrogen.

What is abundant in nature is hydrogen *combined* with other elements – for example, with the oxygen in the water molecule. Often we read in newspapers that *water will be the coal of the future* and, just as often, this sentence is accompanied by a quote from Jules Verne's *The Mysterious Island*:

> *And when the reserves of coal are finished, where will man find the energy necessary to operate its machines? From water. I believe that one day water will be used as a fuel and that hydrogen and oxygen, that make up water, used either separately or together, will provide an inexhaustible source of heat and light.*

Water cannot be compared even remotely to coal, as common experience confirms. Unlike coal, water does not feed a fire, but extinguishes it. Water does not *burn*, because it is already *burned*. To burn means to combine a substance with oxygen – the hydrogen of the water is already combined with oxygen.

Some journalists, however, do not resign themselves. Even if they conceded that there is no hydrogen they would nonetheless argue that it can easily be extracted from water. It's nothing like that at all, because to generate hydrogen from water electrochemically by the electrolytic process requires spending energy (reaction 4):

$$H_2O + \text{energy} \rightarrow H_2 + 1/2\, O_2 \tag{4}$$

This is exactly the same amount of energy that hydrogen can generate as heat when it burns with oxygen to give water. In conclusion, molecular hydrogen is *not* a primary source of energy, for the simple fact that there is none free on Earth. If we want to use it we must first produce it by consuming energy.

We can't even claim that *hydrogen is clean*. In fact, whether it is *clean* or *dirty* depends on the source of energy used to produce it. Using hydrogen as a fuel produced from methane offers no advantage with regard to the environmental impact, since the process involves generating the same amount of carbon dioxide that is produced by burning methane. Similarly, we will have to face all the problems related to the use of nuclear energy if we are to produce hydrogen from this energy source.

The perspectives change completely if we can find a way to produce hydrogen from water using a source of abundant, renewable and non-polluting energy, such as solar energy. Hydrogen can be produced by electrolysis with the electricity generated from photovoltaic panels – at current costs, this is not a cheap way to do it – hydrogen produced by the artificial photosynthesis method is still at the level of preliminary studies. The transition from the present economy based on fossil fuel energy to a hydrogen-based economy, therefore, requires decisive progress in the methods of converting solar energy.

In any case, hydrogen is not a primary energy source, and only when it is economically produced can it be used as an *energy vector*, but not without first resolving other issues related to the matter that hydrogen is difficult to transport, to store, and to use. Therefore it is premature at this time and even counterproductive to place such an emphasis on the hydrogen economy and on the creation of expensive demonstrable prototypes of hydrogen vehicles that car manufacturers exhibit at International Auto Exhibitions.

A great advantage of hydrogen as an energy vector lies in the fact that it can be interchanged directly with another major carrier already widely used: electricity. For instance, with electricity we can produce hydrogen and oxygen by electrolysis of water and, *vice versa*, using devices called *fuel cells* we can generate electricity from hydrogen and oxygen. Without prejudice, however, anyone who wishes to use hydrogen or electricity must first produce it.

The above notwithstanding, even though it will still take some time, production of hydrogen using solar energy remains one of the better solutions to the problem of producing a clean non-polluting fuel for the transportation sector.

8
Energy from Air, Water, and Land

The country that will develop renewables will be the leading nation of the twenty-first century.

Barack Obama

It was quite a windy day in Spain on March 31, 2007. The news from weather forecasts would normally have had no meaning, except that on that spring day the Spanish production of electricity by wind power exceeded, albeit slightly, the energy produced from both nuclear and thermal power. Such an event had never been seen before – it was the first time in Spain and elsewhere. Since then, it has occurred with an increasing frequency.

Wind Changes

The production of electricity by wind power has witnessed an impressive development in recent years. In the course of 2011, the world's total power from this source increased from 198 000 to 238 000 MW (see Figure 23).

Many of the installations are in Europe, providing a total of 94 000 MW with 29 000 MW in Germany alone. However, 26% of the world's energy supply from wind power is found in China, now a world leader in this sector.

New wind power installations built annually in Europe in the last several years have exceeded those of all other technologies. In the mean time, electricity production from coal has stagnated, and electricity produced from nuclear power has been declining.

Europe aims to reach a power level of 250 000 MW from wind power by 2020 so as to meet 12% of the continent's need for electricity and to make a significant contribution to the ambitious objective of producing, by that date, 20% of energy needed from renewable sources.

The production of wind energy is one of the greatest innovations in the energy field in the last thirty years – in view of its magnitude we can no longer treat it as an alternative energy. Rather, it should now be seen as a conventional energy source.

Before installing a set of wind turbines (so-called *wind farms*), it's important to select an appropriate site for such an installation. Several maps are now available

Powering Planet Earth: Energy Solutions for the Future, First Edition. Nicola Armaroli, Vincenzo Balzani, and Nick Serpone.
© 2013 Wiley-VCH Verlag GmbH & Co. KGaA. Published 2013 by Wiley-VCH Verlag GmbH & Co. KGaA.

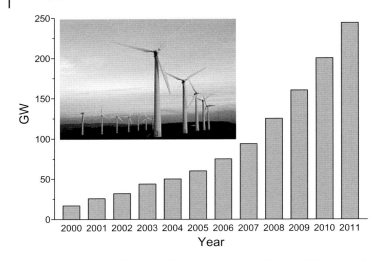

Figure 23 Histogram illustrating the power developed from wind farms installed in various parts of the world between 2000 and 2011. Data source: Global Wind Energy Council, 2012. Photo: Soca/Shutterstock.

with estimates of wind power all over the world. The most favorable sites are found along the European coasts facing the North Sea, in the southern part of South America, in Tasmania (Australia), in the Great Lakes region of North America, and in the Great Plains between the United States and Canada.

More importantly than being strong, the winds needed should be of constant intensity and direction, with an optimum speed around 7 meters per second (m/s). It is estimated that regions in which the annual average wind speed exceeds 7 m/s at an altitude of 80 meters above ground can potentially generate 70 TW – that is, five times the present day average global demand of energy (not only electric!). This represents an immense potential, which may never be completely exploited, but will certainly contribute in a significant way to the transformation of how electrical energy is produced.

On a daily basis, this primary energy is intermittent and seasonal. Transmission and distribution networks connected to wind farms must therefore be prepared for an intermittent electricity flux, typically of medium voltage. Unfortunately, distribution networks in developed countries are not currently conceived to access this energy. They have to use electricity with a predictable and controlled flux selected from a few installations that provide significant power levels. Transition to a massive production of electricity from many small installations, and not just wind farms, will require appropriate and expensive changes to the electrical distribution network.

Since the source of energy is intermittent, building 100-MW wind farms does not mean they will deliver 100 MW of power. The effective annual capacity is about 45% of the nominal capacity in the more windy areas, reaching an average of 30% at the global level. In other words, to obtain 100 MW energy requires wind power farms that can potentially deliver a capacity of 250 MW. It should also be said, however, that no electrical production system operates 100% of the available time, owing to interruptions for maintenance, breakdowns, and other factors.

The problem of intermittency can be mitigated by increasing the reliability of weather forecasts and the expansion of production sites. The greater the number and distribution of the largest wind farms connected to the network, the greater will the stability of the system be, because the average wind distribution will tend to be more homogeneous, thereby moderating the impact of local variations. Accordingly, plans are being developed for a pan-European wind power network linking offshore installations of the Baltic Sea, the North Sea, the English Channel, and an area from the Atlantic to the Western Mediterranean sea, passing through the Iberian Peninsula.

In addition to the intermittency, we often hear talk of other problems attributed to wind farms: noise and the possible impact of the moving blades on birds. In this regard, the question of noise is a problem that has already been addressed and resolved – the latest developments in technology have made these systems more silent than the hissing sound of the wind.

As for the impact on birds, authoritative studies demonstrate that the risk is negligible for almost all species with the possible exception of bats. Moreover, it is estimated that every year hundreds of millions of birds lose their lives through impact with moving vehicles, buildings, and high-voltage power lines. The spread of modern wind farms will not significantly change these figures.

The few disadvantages of wind power are balanced by several advantages. A 10-MW wind farm, which would be sufficient for the electricity needs of 4000 average European families, can be built in less than two months. In similar short periods of time, a wind farm can also be transferred to another site with unprecedented simplicity in comparison to other electrical production facilities. And to upgrade a wind farm it is not necessary to widen it. It suffices to replace the existing blades or otherwise re-install the turbines elsewhere with more powerful blades.

A modern wind farm requires a minimum of maintenance, and when decommissioned the materials used can be recycled almost completely. In just a few months, a wind farm can pay back the energy invested to build it, which distinguishes it among all electrical generation technologies in terms of *payback time*. However, a wind farm may lead to diminished use of the land, although agriculture can continue normally at the sites upon which the turbines are installed. In addition, wind turbines do not need water for cooling, and thus cause no thermal pollution to the environment.

Wind Farms

The cost of health externalities associated with the production of wind energy is estimated at 0.2 Euro cents per kWh in Germany, whereas the corresponding costs of electricity production from coal and gas are, respectively, 30 and 15 times higher.

Table 8 Ten world's largest onshore wind farms as of August 2012.

Wind farm	Current capacity (MW)	Country
Jaisalmer Wind Park	1 064	India
Alta (Oak Creek-Mojave)	1 020	USA
Roscoe Wind Farm	782	USA
Horse Hollow Wind Energy Center	734	USA
Capricorn Ridge Wind Farm	663	USA
Fowler Ridge Wind Farm	600	USA
Sweetwater Wind Farm	585	USA
Buffalo Gap Wind Farm	523	USA
Cedar Creek Wind Farm	551	USA
Meadow Lake Wind Farm	500	USA

Source: http://en.wikipedia.org/wiki/Wind_farm.

In 2010, deployment of wind farms in Europe avoided the emission of about 120 million tons of CO_2 into the atmosphere, equivalent to more than a quarter of the emissions of all cars in Europe.

With regard to security issues to the community, the problem in this case does not even arise. It is unlikely that, even with the worst intentions, someone could devise terrorist actions against wind farms that might endanger public safety.

The world's first wind farm – consisting of 20 wind turbines rated at 30 kW each – was built on the shoulder of Crotched Mountain in southern New Hampshire in December 1980. Most of the largest operating *onshore* wind farms are located in the United States (see Table 8).

As of July 2012, the Jaisalmer Wind Park (India) was the largest onshore wind farm in the world at 1064 MW, followed by the Alta (Oak Creek-Mojave, USA) Wind Farm (1020 MW). The largest proposed wind farm project is the 20 000 MW Gansu Wind Farm in China. Falling costs of wind farms mean that the average onshore wind farm will be competitive with natural gas-fired power generation by 2016.

Europe is the leader in *offshore* wind energy (Table 9) – the first offshore wind farm was built in Denmark in 1991. As of 2011, there were 53 offshore wind farms in waters off Belgium, Denmark, Finland, Germany, Ireland, the Netherlands, Norway, Sweden, Portugal and the United Kingdom – their combined operating capacity is 3813 MW.

More than 100 GW (or 100 000 MW) of offshore projects have been proposed or are otherwise under development in Europe. The European Wind Energy Association expects to install a facility of 40 GW by 2020 and 150 GW by 2030. As of July

Table 9 Ten of the world's largest offshore wind farms as of August 2012.

Wind farm	Capacity (MW)	Country
Walney	367	UK
Thanet	300	UK
Thorntonbank Phases 1 & 2	215	Belgium
Horns Rev II	209	Denmark
Rodsand II	207	Denmark
Chenjiagang (Jiangsu)	201	China
Lynn and Inner Dowsing	194	UK
Robin Rigg (Solvay Firth)	180	UK
Gunfleet Sands	172	UK
Nysted (Rødsand I)	166	Denmark

Source: http://en.wikipedia.org/wiki/Wind_farm.

2012, the Walney Offshore Wind Project in the United Kingdom was the largest offshore wind farm in the world rated at 367 MW, followed by the Thanet Wind Farm at 300 MW also in the UK.

Offshore wind turbines are less obtrusive than turbines on land, as their apparent size and noise are mitigated by distance. Because water has less surface roughness than land (especially deeper water), the average wind speed is usually considerably higher over open water. Capacity factors (utilization rates) are also considerably higher than for onshore locations.

The province of Ontario in Canada is pursuing several proposed locations in the Great Lakes, including the suspended Trillium Power Wind I, which is located about 20 km from shore and over 400 MW in size. Other Canadian projects include one on the Pacific West Coast (see Chapter 11). As of 2010, there are no offshore wind farms in the United States. However, projects are under development in wind-rich areas of the East Coast, Great Lakes, and Pacific Coast.

The European wind industry is the most advanced in the world and employ over 190 000 people. The European Commission estimated that in 2020 the work force in the wind power sector will increase to more than 450 000 – in addition, 2.8 million new jobs are expected to be created across Europe through the expansion of wind power. At the end of 2011, about 80 000 people were employed in the United States wind industry.

After years of troublesome stalemate, Italy is trying to recover lost ground in this industry, which has a huge potential for growth in the coming decades. At the end of 2011, the installed capacity in Italy amounted to 6940 MW, placing it seventh in the world. In 2011, Italian wind farms produced 9860 GWh, equal to 3.1% of national electrical consumption or 14% of residential demand. For a country that does not have large wind resources this result is encouraging.

The esthetic impact of wind farms is one of the factors that have slowed their installation in Italy. Certainly, it is an aspect to be taken into account, but we believe that the criticism and protests are often spurious. Italy is littered with abuses of every kind – so much so that in the countryside it is almost impossible to take a photograph that does not include high-voltage transmission lines, microwave relay stations, telephone antennas, or some other technological wonder. To be sure, it is not these modern windmills that will decisively deface the landscape. But then, taste may change. Maybe one day we may find the presence of these wind farms non-obtrusive and esthetically acceptable (see Chapter 11), especially in light of the substantial accrued benefits such as a positive impact on climate and on people's health.

In this regard, in a study published on October 2010 in the scientific journal *Proceedings of the National Academy of Sciences* (PNAS), Somnath Baidya Roy, a professor of atmospheric science at the University of Illinois, showed that in the immediate vicinity of wind farms, the climate is cooler during the day and slightly warmer during the night than the surrounding areas. According to Roy, the effect is due to the turbulence generated by the blades.

In another study, presented at a San Francisco conference in December 2010, Gene Takle and Julie Lundquist of the University of Colorado noted that their analysis, carried out on corn and soybean crops in the central areas of the United States, showed that wind turbines generated a microclimate that in fact improved crops as it prevented formation of Spring and Autumn frosts and reduced the action of pathogenic fungi that grow on leaves. Even at the height of summer heat, a lowering of 2.5–3.0 degrees above the crops was observed due to the turbulence caused by the blades – this made a significant difference in the cultivation of maize.

Compared to the environmental impact of traditional energy sources, that of wind power is relatively minor, as it consumes no fuel and emits no air pollution, unlike fossil fuel power sources. The energy consumed to manufacture and transport the materials used to build a wind power farm is equaled by the new energy produced by the farm within a few months. While a wind farm may cover a large area of land, many land uses – such as agriculture – are compatible. Only small areas of turbine foundations and infrastructure could make land use unavailable.

Present-day countries that are leaders in wind energy are China and the United States, with the leaders in Europe being Denmark, Spain, and Germany, in which wind power covered, respectively, 26%, 16% and 10.6% of their national electricity needs in 2011. With their wind farms, these countries dominate the world's market in the production of wind turbines.

In the early 1980s, a typical wind turbine had a diameter of about 15 meters and delivered an electrical power of 50 kW. Today there are blade models 125 meters in diameter capable of delivering a power of 6000 kW (6 MW). Ten-MW blades for wind farms at sea are in the design stage – they are to be positioned along shallow coastlines.

Power from a wind turbine has increased more than 100 times in the last twenty-year period, which has led the costs of electricity production to decline by about 80%. Clearly then, the price of energy from wind power is competitive with the cost of electricity generated from thermal power plants. While the cost of oil increased at least three-fold and that of uranium more than 10 times in the last decade, the wind costs nothing.

Locations of available resources are known with great precision in every corner of the Earth. Thanks to technological advances, therefore, the cost of wind electricity can only decrease. This is why international financial groups are investing heavily in wind power and other renewable energy sources, rather than in the more traditional sources of fossil fuels and nuclear power, both of which are burdened by large uncertainties.

Water–between Past and Future

Seventy years ago, Italy was a virtuous country in terms of energy – electricity came almost exclusively from renewable sources. Of the 15.5 GWh produced in 1938, about 14.6 GWh came from what the propaganda of the time called *white coal* – that is, from national hydroelectric power originating from dams situated in the Italian Alps and in the Apennine mountain range. This led inexperienced advocates of self-sufficiency of the 1930s to propose such outlandish (maybe a bit crazy) projects as electric plowing. Nonetheless, the fact remains that exploitation of waterfalls to generate mechanical energy or electricity was of vital importance for the start of Italy's industrialization in the period overlapping the nineteenth and twentieth centuries. For example, the growth of the textile mills in northern Italy and steel mills in Terni (in the Umbria region) were inextricably linked to the availability of this source of energy.

Italy's electrical consumption today stands at about 300 000 GWh – 20 times greater than in 1938. Some of the *white coal-fed* power stations of the 1930s are still in operation. Hydrological resources currently account for 15% of Italy's electricity consumption, a very high fraction for a developed country. By contrast, hydroelectric power accounts for only 4% of Germany's consumption, whereas it is 7% in the United States. In some South American and African countries, hydroelectric power remains the main source of electricity – in Brazil it exceeds 80%.

There are presently 800 000 dams in operation worldwide, a number of which are fairly tall (45 000 being higher than 15 meters). Hydroelectric power currently provides 2.3% of the world's primary energy and 16.2% of its electricity; the latter amount is greater than the quantity of electricity from nuclear energy, at 13.4%.

However, hydroelectric power stations occupy 60% of the surface of the entire global energy infrastructure. Globally, dams presently cover 300000 km², an area about as large as Italy. It is therefore a very invasive technology.

After the boom of the 1960s and 1970s, construction of large hydroelectric power stations subsided significantly because the problems created by this technology were more than expected. Construction of dams has significant consequences for people and for the environment – forced expulsions of large populations from areas that need to be flooded, excavation of huge amounts of material, changes in microclimate, dangers for the people downstream of the dam, depletion of biodiversity, distortions of fluvial wildlife, spread of diseases such as malaria transmitted from pests that thrive in water basins, drastic reduction of the average speed of river flow with rise of pollution, sedimentation of materials at the bottom of reservoirs with progressive loss of electric power output and high maintenance costs, and decreased fertility of the land in the valleys downstream.

Recent studies have shown that in tropical areas hydroelectric dams can become major producers of greenhouse gases (CO_2, CH_4) owing to the decomposition of organic material in the warm and stagnant waters. In some cases, it was calculated that the equivalent production of electricity with thermoelectric facilities would be less damaging from the point of view of production of gases that might alter the climate.

Some of the older hydroelectric facilities built in the first half of the twentieth century around the world are being dismantled. This is a very complex and expensive operation, but otherwise necessary for reasons of safety and environmental restoration. For several years now, the rate of decommissioning large hydroelectric power installations in the United States has surpassed the rate of construction of new ones.

The energy from waterfalls was widely used in Europe and North America, where it is estimated that over 70% of exploitable potential is already operational. In Africa and Asia, hydroelectric power exploitation stands at between 10 and 25%, and, not surprisingly, it is there that we find the few existing construction projects for new large power dams.

Of these, the best known is the Three Gorges Dam on the Yangtze River of China, a pharaonic project that provides this Asian economic giant with 22 500 MW of electrical power (greater than the total capacity installed in Austria).

To realize such a gigantic hydroelectric dam has necessitated the flooding of an area approximately 600 km long and 2 km wide, the dismantling of no less than 23 towns, the mandatory displacement of 1.3 million people, and the wiping out of great archeological sites. With this project, Chinese authorities expect to control the floods that periodically affect the river basin.

However, the worrisome environmental impact of this enterprise is already manifesting itself in various ways – for example, landslides. Chinese authorities are now actively involved in preventing such an important symbolic project as the Yangtze River Dam from becoming an ecological and economic disaster, not to mention the damage to China's image and reputation in the rest of the world.

Unfortunately, the approach to build mega-projects gives but scant attention to a possible negative impact on the environment. Mega-projects are spreading throughout Africa, where Chinese companies are reaping huge financial profits while supporting political regimes that are insensitive to environmental issues and human rights. Impetuous growth of hydroelectric installations has also taken place in the West, dictated by the needs of the post-war economic boom. Such installations have had some disastrous consequences.

On the evening of October 9, 1963, following test operations of the Vajont power dam, near Longarone (in the province of Belluno, Italy), a landslide of some 30 million cubic meters of water and mud from the nearby Toc mountain swept through the Piave river valley, causing nearly 2000 casualties. This immense and avoidable tragedy should remain as a perennial warning to man that he should not be bewitched by a senselessly voracious economic growth with no regard for the laws of nature.

Faced with many such problems, however, we must nonetheless note that hydroelectric power has many merits. First, it is a relatively simple and economically viable technology. Despite the fact that construction times are long and require large capital investments, electricity costs are among the lowest ever. The dams provide an ideal way to store energy for the water basins can be *reloaded* during night hours using excess electricity often imported at low cost so as to have enough water reserve to begin operation the next day within a few seconds during the daily peak demand for electricity.

In addition, hydroelectric dams are often used not only to produce electricity, but also to provide drinking water and water for irrigation. More to the point, these basins constitute a useful tool for the control of water sources, particularly in periods of drought or abundant rainfall.

At the global level, there is no doubt that the spread of hydroelectric power has contributed in a relevant manner to avoiding atmospheric intake of pollutants and greenhouse gases. If the energy currently obtained from waterfalls were produced from thermal power plants, we would have a significant annual increase in the carbon (+15%) and sulfur (+35%) entering the atmosphere.

The prospects for the expansion of hydroelectric power are now focused on *small is beautiful*, that is, on power plants less than 10 MW in size, sometimes even a few kW, which can exploit a small but constant flow of water. This possibility is not only interesting for the rich countries, where we hardly find any new large installations under construction, but also for rural areas that are not connected to the electricity grid in developing countries. Currently, Italy produces 2600 MW power from small hydroelectric installations of less than 10 MW each.

In summary, water power can be stored indefinitely, and is therefore an attractive option to support and integrate renewable technologies that produce energy intermittently such as windmills and photovoltaic devices.

Geothermal Energy

July 4, 1904 is an important date in the history of energy. For the first time on that day five light bulbs were lit by an electric motor powered by steam emitted from the bowels of the Earth. The architect of this brilliant achievement was Prince Piero Ginori Conti, a young Tuscan who brought to fruition a path started long ago by the grandfather of his father-in-law, Francis Larderel.

Larderel, a descendant of a noble French family which moved to Tuscany at the time of Napoleon I, was fascinated by the strange whitish sludge around natural fumaroles that flowed underground on the metal-bearing hills of the province of Pisa. This sludge consisted of boric acid, a substance present since Etruscan times that was put to various uses, particularly in the preparation of glazes and paints. Larderel's family built a large fortune around boric acid. A century later, the family heir, Ginori Conti, taking advantage of those spectacular phenomena of nature, founded a second even larger enterprise to produce energy on a massive scale.

The internal structure of spaceship Earth is characterized by three concentric enclosures of different thicknesses: the *crust*, the *mantle*, and the *core*. The core is the nucleus of our planet, a huge ball of iron and nickel with a radius of about 3500 km and consisting of two parts: the outer part is fluid and has an average temperature of 3000 °C, whereas the temperature of the inner part exceeds 4000 °C but is solid because of the enormous pressure to which it is subjected.

The crust is the outermost layer but thinner, with an average thickness of about 30 km on the continents and 5 km under the oceans. Together with the external part of the mantle, the crust forms the *lithosphere*, which is divided into ten major plates of varying shapes and sizes, which can be compared to rafts floating on the mantle below. Friction at the sites between these plates is the cause of earthquakes.

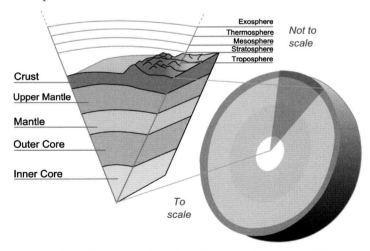

The mantle is the intermediate layer between the crust and the core. It is 2900 km thick and consists of rocks rich in iron and magnesium. These rocks are also subjected to very high pressures and temperatures. Volcanoes are natural chimneys that communicate directly with the surface. The entrails of the Earth are thus very hot. Deep underground the temperature increases by an average 30 °C per kilometer. The endogenous heat of the planet is due mainly to two phenomena: the convective movements of the mantle fluid that redistributes the core's thermal energy outward, and the radioactive decay of isotopes (uranium, thorium and potassium) that are nested within the crust.

The heat of the interior of the planet reaches the Earth's surface with an average power of about 0.06 W/m². Dividing the surface of Italy by the number of its inhabitants shows that every Italian is accompanied by about 5200 m² of the available land from which 2700 kWh of thermal energy could be released each year. It is a sizeable amount, equal to about 3 to 4 times the average annual power consumption of every Italian.

In some areas of the Earth, the underground temperature is rather high even at relatively low depths. This is indeed the case in the valleys of Tuscany, which have witnessed the work of pioneers who tried to exploit geothermal energy. The Larderel field, which takes its name from its founder, occupies about 200 km², with the underground temperature exceeding 300 °C at a depth of only 3000 meters.

The presence of hot rocks at low depths is a necessary but not sufficient condition for obtaining exploitable energy from the Earth's depths. It is also essential that these heat *bubbles* come into contact with rainwater that has penetrated deep into the Earth through fissures and permeable rocks – a favorable conformation of the Earth's crust. This creates the conditions to generate the fluid media (hot water and steam) used to produce heat and electricity on a large scale. These aqueous media are often constrained underground such that their extraction and exploitation necessitate drilling suitable wells using technologies similar to those used for extracting oil and gas. Sometimes, however, hot water and steam find a path that generates spectacular phenomena, at times pleasant and healing, such as hot springs, fumaroles and geysers.

Depending on specific local conditions, in particular morphology and the subsurface temperature, geological reservoirs are formed that consist mostly of *hot water* or mostly of *steam*. The first are far more frequent. Here water can reach temperatures above 300 °C, though it remains a liquid due to the prevailing high pressures. When the dig of the artificial well reaches these reservoirs, the water rises to the surface, turning partly into steam (lower pressure as it reaches the surface), which can then be used to turn the turbines that generate electricity.

Deposits dominated by steam are the most valuable because the dry steam is practically ready to be inputted into the turbines. The Larderel reservoir is of this type and is exploited to generate a total of 700 MW of electrical power, similar to that of a conventional power facility fed by fossil fuels or by nuclear power. The largest geothermal field in the world, *The Geysers* in California, produces nearly 800 MW of electricity (see Chapter 12).

Worldwide, geothermal electric power installations account for a total of approximately 10 700 MW. They are located mostly in Italy, the United States, Japan, the

Philippines, Central America, and Iceland. This source of energy provides approximately 0.3% of the electricity needs of the planet at very competitive costs compared to traditional sources.

The geothermal sector is predicted to grow in view of its enormous potential. A study by the Massachusetts Institute of Technology (MIT) in Cambridge (a Boston suburb, USA) calculated that the geothermal reserves of the United States at a depth of about 10 km are 130 000 times the current annual energy consumption of primary energy in the country. However, taking into account the economic, technical, and environmental constraints, the study predicted that in 2050 the geothermal energy will provide only 10% of the electricity needs of the United States, a quantity that could, nonetheless, make a significant contribution to the transition of energy from conventional to renewable sources.

Geothermal resources are renewable, but only if well managed. Until the 1970s deposits were simply exploited, while today the trend is focused on growth. To avoid the rate of extraction of geothermal fluids exceeding the ability of natural replenishment from rainfalls, water is re-injected into the geothermal reservoirs. This is a complex operation from a technical point of view, but one that makes it possible to revitalize geothermal fields that have shown signs of dwindling.

Of course, this technology presents some problems, the most trivial of which is related to the extraordinary sensitivity of our nose to sulfurous substances, which can be detected in infinitesimal concentrations. The hydrogen sulfide (H_2S) generated from the decomposition of organic wastes in the subsoil is the cause of the unpleasant smell of rotten eggs that often accompanies geothermal phenomena. Other more substantial critical points have to do with the sinking of the land (subsidence), the chemical aggressiveness of the geothermal fluids, and the visible and noisy impact of large installations.

All the above problems have been largely resolved or otherwise greatly reduced. At present, the main criticism of the technology remains the high cost of exploration and development of geothermal fields, which can represent up to two thirds of the total cost, sometimes without leading to concrete results.

Another problem is strictly related to thermodynamics: geothermal energy production converts low-quality energy (heat) into high-quality energy (electricity). Accordingly, the conversion yield tends to be rather low. In Italy's geothermal fields, with steam at an average temperature of 220 °C and pressure of 10 atmospheres, the electricity conversion efficiency does not exceed 20%.

The best-known use of geothermal energy is linked to the production of electricity. However, geothermal heat is also employed directly in industrial and residential sectors that need sources of relatively low temperatures (30–150 °C) – for example, in heating fish farms, swimming pools, domestic dwellings and greenhouses, and in drying lumber and agricultural products.

In the Iceland capital, Reykjavik, the substantial thermal requirement of 160 000 people is entirely covered by geothermal energy. It is the largest heating system in the world. Overall today, the direct use of endogenous heat of the Earth provides a thermal capacity of about 50 000 MW.

In recent years, the use of geothermal *heat pumps* has experienced significant growth. Heat pumps are electrical devices that exchange heat with the ground or with the aquifer in the vicinity of buildings. They exploit the fact that at a depth of a few meters, the temperature of the land remains virtually constant throughout the year, as anyone who owns a good cellar can attest to. During the winter months, the soil has a higher temperature than the outside air, while in the summer the situation is reversed. Hence, it's possible to have throughout the year a natural reservoir that can heat or cool buildings with a much reduced consumption of electricity.

Geothermics is a geophysical science that is very ancient and at the same time very advanced. It needs the support of several branches of chemistry, physics, and engineering. Italy is at the forefront of the world in this area, thanks to an ample experience in the field. Research and development projects today focus on so-called EGT (enhanced geothermal activity), namely the extraction of heat from the ground down to 10 000 meters (today we can reach at most 5000 meters) in areas of low permeability and low porosity. This would open up great potentials for using geothermal energy worldwide, and not just in those very small areas where nature has made it easy to exploit this very valuable energy source.

Sea Power

The snowfields of a beautiful alpine wintry landscape constitute a form of potential energy which, in the warmer months, can be released and used, in part, to feed hydroelectric facilities. Seasonal thaw is a process sufficiently slow and progressive, so much so that it can be harnessed and exploited for useful purposes. However, when the snow makes its way down a valley as an avalanche, there is no possibility of harnessing all the energy dissipated in just a few moments.

Many natural events of short duration are associated with the release of large amounts of energy. Take, for example, lightning, which can develop an electromagnetic power of one trillion watts (1 TW). Man-made satellites record worldwide an average of 1.4 billion lightning strikes annually, about 45 every second. Some people suggest that we should exploit these natural phenomena or others even more powerful, such as volcanic eruptions, earthquakes, tropical hurricanes, or, more modestly, thunderstorms and hailstorms, for energy purposes.

In practice, however, the only unconventional renewable resources that can be effectively used are those associated with the kinetic energy of waves at sea, a phenomenon often inconspicuous but continuous. Water movements are associated with the action of gravitational forces (tides), the generation of surface disturbances by wind action (waves), and with temperature or density differences between the surface layers and those layers deep in the oceans (ocean currents). The total power of oceanic currents is estimated at 100 GW. Given the vastness of the surface of the seas on our planet, it represents a modest power of $0.3 \, mW/m^2$. This form of energy can become significant in narrow straits because the flow speed increases with decreasing flow width. Sailors in ancient times were well aware of this when crossing the Strait of Messina between Scilla and Cariddi in Sicily.

From a quantitative point of view, however, exploitation of tides is much more promising, as they move huge masses of water in relatively short periods of time. It is estimated that the exploitable potential of the gravitational energy of the tides is at least 360 GW on the global scale. Of course, exploitation of these phenomena should be concentrated in areas where their potential is much greater than observed in the Mediterranean Sea: for example, in some coastal areas of Canada and Alaska, in the southern part of Argentina, and in France. In these regions, differences in tidal waters range from 5 to 15 meters and masses of fluid in motion are enormous. Exploitation of these phenomena is comparable to that of major rivers with small waterfalls.

The only power station in the world that provides significant power (240 MW) from tidal waves is the one located in Saint-Malo in Northern France. Electricity production is done using both flow directions. A serious problem for systems of this size is their position – the dam effectively shuts down a stretch of the coastline and makes navigation very difficult. For this reason, experiments are being carried out with some success with smaller submerged installations, which exploit the underwater currents generated by the tides. These installations are substantially similar to the blades of wind farms, but are considerably smaller in size. One of the key challenges for these technologies is the resistance of materials to the corrosive effect of seawater. On the other hand, a very interesting aspect of the exploitation of tides is the extreme precision, virtually unmatched for a renewable source, for which we know their timing and the amount of energy that can be delivered to an electricity network.

Marine waves are systems of concentration of wind energy, thanks to the higher density of water compared to that of air. The estimated wave power of the seas is immense, approximately 90 TW (remember that the world demand for power today stands at around 16 TW). The first patent application to exploit the energy of sea waves was deposited in France in 1799. The interest in this never dormant technology resurfaces regularly when energy crises materialize, as occurred in the 1970s and is occurring even as we speak.

Because of wind behavior across the globe, among the more potentially interesting areas are the western coasts of the continents in the Northern Hemisphere at the mid-latitudes: for example, Portugal, Scotland, and California. There are dozens of pilot projects currently being examined off these regions that attempt to exploit various types of machines to harness sea power. A few of these will likely reach commercialization.

The major technical challenge is the resistance of these machines to corrosion, so that they must be designed to withstand adverse weather conditions far more severe than usual, because a single particularly intense atmospheric event can destroy a whole facility. Among the opponents of these projects are the fishermen and the wind-surfers; however, it should not be very difficult to find space at sea for all these folks. If this technology were to find a sufficient number of investors, it might find its own interesting niche for development. A recent report estimated that the United Kingdom could get as much as 20 000 MW of electrical power from sea waves by the middle of the twenty-first century.

Another form of marine energy which has been under examination for some time involves the exploitation of the oceans' temperature gradients at different depths to produce electricity. A large number of financial resources were invested in this area in the 1970s but with modest success.

The technology, known as OTEC (*ocean thermal energy conversion*), requires a thermal jump of at least 20 °C between the sea surface and a depth that does not exceed 1000 meters. This makes those ocean areas at the equator particularly attractive, especially those in the Western Pacific, although so far there are only demonstration-type systems. More promising is the thermal use of deep sea water basins in the proximity of large cities. For instance, a centralized air conditioning system serves some skyscrapers of the financial district of Toronto, Canada, which takes advantage of the cold waters of Lake Ontario. The electrical power savings exceed 7 MW.

Renewable hydrological technologies that we have cited are still, in many cases, at the initial testing phase. It should be noted, however, that even wind energy, which today is experiencing unparalleled growth, was in a similar situation only twenty years ago.

In a world in search of clean energy resources and with the necessity of disengaging from fossil fuels, it is reasonable to expect that these energy technologies, today *minor*, will likely be part of the renewable energy mix that will be put in place during the twenty-first century. Their eventual success will, of course, depend heavily on the availability of funds for scientific and technological research. These have so far been somewhat limited when compared with funds expended on other energy technologies.

9
Fukushima and the Future of Nuclear Energy

Reality has a disconcerting habit of putting us in front of the unexpected for which, in fact, we were not prepared.

Hannah Arendt

At 2:46 PM on March 11, 2011, local time, an earthquake of magnitude 9.0 on the Richter scale, with its epicenter at sea, hit the North-Eastern coast of Japan. Together with the tsunami that followed, the earthquake caused severe damage to 11 nuclear reactors located in 4 different nuclear installations. The most serious consequences occurred at the Fukushima-1 (Fukushima Daiichi) nuclear facility, which included six nuclear reactors. The facility was run and managed by the Tokyo Electric Power Company (Tepco).

What Happened at Fukushima Daiichi?

Reactors 1, 2, and 3 were operational at the time of the earthquake, while reactors 4, 5, and 6 were turned off for maintenance. Reactor 4 had been emptied earlier of all its fuel rods – they had been placed in the cooling bath.

Following the earthquake, reactors 1, 2, and 3 stopped automatically. However, nuclear reactors continue to generate heat from the spontaneous radioactive processes, even though the chain reaction has been blocked. Accordingly, it was absolutely imperative for cooling to continue. The earthquake also damaged the electric grid pylons, causing a *blackout*. Soon after, the diesel engines came into emergency action to operate the water pumps. After less than an hour, a tsunami with 14-meter waves, more than double the height of the protective wall of the plant, hit the facility and completely flooded the pumps, rendering the diesel engines unusable.

For some time, partial pumping of cooling water continued with energy supplied by an emergency battery. Unfortunately, cooling of the reactors failed, and the reactors began to overheat.

Later, because of a lack of cooling, the water in the reactors and in the pools partially evaporated, leaving the fuel rods exposed. Temperature increased further to the point when the water in contact with the overheated materials (in particular,

Powering Planet Earth: Energy Solutions for the Future, First Edition. Nicola Armaroli, Vincenzo Balzani, and Nick Serpone.
© 2013 Wiley-VCH Verlag GmbH & Co. KGaA. Published 2013 by Wiley-VCH Verlag GmbH & Co. KGaA.

Figure 24 A satellite photograph of the four reactors at the Fukushima Daiichi nuclear facility damaged by explosions in March 2011. (Reproduced by courtesy of Digital Globe).

zirconium used to cover the fuel rods) underwent thermal splitting, generating hydrogen that ultimately exploded and destroyed the upper portions of the buildings that housed reactors 1, 3, and 4 (Figure 24).

As the temperature climbed, the more volatile fission products, iodine-131 (^{131}I) and cesium-137 (^{137}Cs) were discharged into the atmosphere. In attempting to limit the temperature rise, technicians poured sea water on the reactors from helicopters, a measure that proved completely ineffective because the water was dispersed by strong winds. Then, water cannons were used to douse the reactors. This only caused a strong flow of radioactive materials to be discharged into the ocean.

As a result of the increase in temperature, all the fissionable fuel of reactor 1 – and probably also that of reactors 2 and 3 – was liquefied together with part of the concrete structures, creating a highly corrosive and radioactive *magma* (a mixture of molten rock, volatiles, and solids), which reached temperatures as high as 2500 °C.

What the real situation is inside the various reactors will not be established for months, if not years, to come. At the Three Mile Island facility in the USA, technicians had to wait three years after the accident before they could inspect the reactor core with a camera.

The Fukushima disaster, initially classified as a level 4 accident in the INES scale (International Nuclear and Radiological Event Scale), was subsequently raised to the maximum level 7, the same level as for the Chernobyl disaster.

The Japanese nuclear safety agency estimated that the radioactive material discharged into the atmosphere in the first month after the accident was approximately 10% of that emitted in the Chernobyl accident. No doubt, significant quantities of radioactive material were also released afterwards, particularly into the ocean and into the soil around and under the reactors. Tepco estimated the time needed to complete the cooling of the fuel rods remaining in the reactors and in the water pools to be six months. Once the situation is back under control (and who knows when that will be from the point of view of safety), only then will it be possible to decide what to do next. In any case, the problem will certainly continue for decades to come, as the Chernobyl experience has taught us.

The Consequences for the Population

The release of radioactive material from nuclear reactors 1, 2, and 3, and from the water pool of reactor 4 has continued for months. Diffusion of radioactive waste into the air and the subsequent impact on the ground has particularly affected the area north of the nuclear facility. Fortunately, after the accident the wind direction was mostly to the East toward the ocean.

The two main radioactive elements emitted by the incident – ^{131}I (half-life, 8 days) and ^{137}Cs (half-life, 30 years) – were subsequently detected worldwide, albeit in small quantities.

Japanese authorities admitted later that the Fukushima incident also released strontium-90 (^{90}Sr; half-life: 28 years) and small amounts of plutonium-239 (^{239}Pu; half-life: 24000 years), presumably from reactor 3 that was powered by MOX (*mixed oxide*), a fuel that contains 5% plutonium (see below).

The unit of measure of a radiation dose absorbed by a person is the sievert (given the symbol: Sv). The threshold maximum recommended by international authorities for workers in nuclear facilities has been set at 20 millisieverts (mSv) per year. The Japanese Government had set the dose at 100 mSv/year; it was later raised to 250 mSv per year after the accident. At least 50 workers were contaminated by the radiation above the latter dose limit.

The distance at which the population might be in danger of radiation contamination was extended from an initial 3 km to 10 km around the nuclear facility, and then raised to 20 km and then finally to 30 km. In total, the number of people evacuated was at least 80000, many of whom received unspecified doses of radiation. The information given by Tepco and the Japanese authorities was always scanty, so much so that it even raised protests from many other countries. Radioactive contamination has affected the air, water, vegetables, meat, and fish. Farming and fishing have been prohibited in the evacuated zones. Also discouraged is the consumption of tea, a drink preferred by the Japanese people – tea exports have also been prohibited.

It is difficult to determine what the real risk to human health will be. Exposure to radiation can cause cancer. However, given the long latency of these diseases, it will be very difficult to distinguish the effects of the nuclear accident from those with other causes. If a serious epidemiological study were to be carried out after a nuclear accident (something that was not done after the Chernobyl accident), it would be only after decades that appropriate statistical estimates could be made, estimates that would carry much uncertainty.

It suffices to think of the many that have died or will die prematurely as a result of the Chernobyl accident. According to the United Nations Scientific Committee on the Effects of Atomic Radiation (UNSCEAR), the number of victims will likely amount to between 65 and 4000 in about 80 years. Other estimates from the former Soviet Union and the National Academy of Sciences (NSF) of the United States predict the casualties to reach 1 million; Greenpeace even puts this number at 6 millions. It must be said that the members of the UNSCEAR Committee, who minimized the damage caused by the Chernobyl accident, if not controlled by the nuclear lobby certainly has strong connections to it.

The discrepancies between the various sources are due to various reasons. Rarely does exposure to radiation show immediate lethal effects, as happens in other accidents. If the people that were affected by radiation (often not well known) die of cancer after many years, it will be impossible to demonstrate direct evidence that cancer was caused by the radiation.

We must then consider that radiations have effects that go beyond the physical damage. People evacuated will be forced to live for a long time away from their homes, tormented by the fear that they have absorbed doses of radiation sufficient to jeopardize their health – they live under the threat of a time bomb that can blow up at any time. As already witnessed at Chernobyl, the evacuees will be easy prey to depressive syndromes that can lead to greater vulnerability to other diseases, to alcoholism, and even to suicide.

A Lesson from the Fukushima Disaster

Fukushima has confirmed what some people have suspected for a long time, and with good reason: absolute nuclear safety does not exist – the unpredictable cannot be predicted. If we wished to increase security, we would have to increase the complexity and the robustness of the nuclear installations, which can only lead to increased costs.

Fukushima has also confirmed that a serious nuclear accident, unlike any other, is not definable in space or time. Radioactivity is in fact transmitted largely through the atmosphere and the food chain, neither of which we can control, and land use can be compromised for thousands of years. For these very reasons, no insurance company covers damage resulting from a serious nuclear accident – even Governments cannot cope with such eventualities. At the same time, the Fukushima incident also confirmed that damage decreases with distance from the accident site. Accordingly, speeches by those in Italy who argue that there are many nuclear power installations beyond the Alps which, in the event of an accident, would cause as much damage as if they were here in the Po Valley are totally meaningless.

Fukushima also teaches us – and even this has been known for some time – that the enormous economic and political interests of the nuclear establishment preclude a transparent management of nuclear accidents. So how can we trust what we're told by the companies that manage nuclear facilities and even more so by our elected officials? In this regard, the Tepco Company that managed Fukushima has been known for some time to have falsified security data. Yet it was only with coercion that the Japanese Government was able to take over the situation, albeit a few days after the incident. In turn, the Government failed to consult the special crisis unit and failed to consider the data collected by the existing national network of dosimeters – instruments that measure the intensity of ionizing radiation.

It should be added that it is still not clear what the role of the International Atomic Energy Agency (IAEA) might be in the case of nuclear accidents, nor is its degree of independence from member States. For many days after Fukushima, the IAEA did nothing except to transmit whatever news was given by Tepco and the Government of Japan. Only later, two months and ten days after the incident, did the IAEA send a group of experts to Japan to investigate the situation.

Fukushima also teaches us that a serious nuclear accident can not only cause an economic collapse[1] of the Company that manages a nuclear facility, but can also cause many problems for the whole country, especially for a country that depends strongly on nuclear energy, such as Japan.

After the Fukushima incident, only 16 of the 54 Japanese nuclear reactors remained in operation for some time. Industrial production fell heavily as a result of electricity outages. Lifestyles had to change. The use of elevators and air conditioners was drastically limited. People had to resort to opening their office windows. Lighting in shops and stores was reduced. Employees went to work without jacket and tie. Maybe if people had thought of consuming less energy, a smaller number of nuclear facilities would have been constructed.

Another no less important lesson Fukushima teaches us is that in its attempts to make greater profits, nuclear technology, already dangerous in itself, can increase the risks to the population.

MOX,[2] a fuel which consists of a mix of uranium and plutonium obtained from spent fuel, and which fueled (in part) reactor 3, is far more dangerous than just uranium. France and the United Kingdom, unlike the United States, are very active in the field of reprocessing fuel, and thus had been contracted by Japan to recycle Japan's exhausted fissionable material and provide it with the fuel MOX. Now that Japan and other nations have abandoned plans to extend the use of MOX, France and the United Kingdom find themselves in some economic difficulty. In particular, the British have closed an expensive facility that produced MOX, which was built in Sellafield and never really entered into operation. A secret message to the United States Embassy in London, and divulged by Wikileaks, talks about it being the most embarrassing economic disaster in British

1) Tentative estimated damage: 100 to 200 billion Euros, equal to the cost of constructing 30 to 50 nuclear facilities; for comparison, the compensation fund imposed by the US Government on British Petroleum for the 2010 disaster in the Gulf of Mexico from the Deepwater Horizon platform was 20 billion dollars.

2) **Mixed oxide**, or **MOX fuel**, is a blend of plutonium and natural or depleted uranium which behaves similarly (though not identically) to the enriched uranium feed for which most nuclear reactors were designed. MOX fuel is an alternative to low enriched uranium (LEU) fuel used in the light water reactors which predominate in nuclear power generation. Some concern has been expressed that used MOX cores will introduce new disposal challenges, though MOX is itself a means to dispose of surplus plutonium by transmutation.

industrial history. Finally, the Fukushima incident has laid bare the danger of idolizing the nuclear technology.

Japan is a country with scarce natural resources. In the post-World War II period, just after the destruction of Hiroshima and Nagasaki by atomic bombs, Japan thought that nuclear energy might offer an ideal solution to resolve the energy problem.

Government pressure and extensive, continuous, and costly advertising campaigns by electrical utilities over the years built the myth that nuclear energy was safe. Indeed, textbooks, public relation agencies, and theme parks—aimed particularly at children—all describe nuclear power as being a Wonderland. They instilled the idea that nuclear energy was not only necessary, but was also absolutely safe.

So it happened that, in a country like Japan, where a car after just three years of use is subjected to detailed verifications for it to be considered roadworthy, decades-old nuclear reactors have been controlled exclusively by those who had no interest in stopping their operation.

What Is Today's Cost of Nuclear Energy?

For various reasons, it is not possible to give a reply to this question. However, it is an indisputable fact that nuclear energy is not cost-effective in a free market economy, which requires that any new proposed nuclear facility be developed, built, and managed by the private sector. In that case then, the private sector should also bear the costs of decommissioning the nuclear reactors and for the management of radioactive wastes.

The rating agency Moody warned that any project designed to build a new nuclear power plant may increase the risk of the construction company seeing its *rating* downgraded. For its part, Citigroup, once the largest financial service provider in the world, has stated categorically: *New nuclear? The economics say no.*

There is no nuclear power plant in operation in the world that has not benefited, in one way or other, from non-repayable State aid in the form of subsidies to production or otherwise direct subsidies to companies involved, or to this day does not continue to get support for waste disposal and decommissioning of nuclear installations at the end of their useful lives.

One of the main unknowns about costs of nuclear energy is the storage of highly radioactive wastes. We saw in Chapter 6 that the United States, a vast country with the most advanced technology in the nuclear sector, abandoned the construction of a permanent storage at Yucca Mountain. At present, the wastes are mostly stored in baths or in tank-containers located in the grounds of the nuclear facilities. China hopes to *select* a suitable site for its nuclear wastes by 2020. The site is expected to be ready—maybe—by 2050.

Nuclear facilities that have come to the end of their useful life are in themselves gigantic nuclear wastes, although certain misleading advertising wants us to believe that the site on which the facility stands can be transformed into a garden. Disposal of nuclear residues is very, very expensive, so much so that even the

Italian Minister of Finance, at a meeting of EU Finance Ministers, stressed that in assessing the economic situation of a State the extent of its *nuclear debt* must also be considered.

Attempting to revive the nuclear energy sector in Europe with the construction of a new *European Pressurized Reactor* (EPR) in Finland, the French company Areva has encountered enormous difficulties. The roadmap was for the work to end in 2009, but that year the company estimated a 4-year delay in construction. We're now looking at 2015 for the entry of this EPR facility into production of electricity. In the meantime, the cost has practically doubled compared to the 3 billion Euros initially estimated.

In the United States, where more than 100 operating nuclear reactors are inexorably aging, construction of new nuclear facilities is postponed or has been abandoned altogether, despite the commitments made by the Bush Administration to take care of the waste problem and despite the financial guarantees granted by the subsequent Obama Administration. Nonetheless, permits have been issued by the Obama Administration for two new nuclear power stations to be built in the USA.

What happened in October 2010 is emblematic. The project to build an EPR facility by UniStar, a consortium composed of Areva and the American Group CEG (Constellation Energy Group) was abandoned due to the withdrawal of CEG from the project. As soon as the news spread, the shares of CEG on the New York Stock Exchange increased in value.

New nuclear installations are currently being built, mostly in countries with a centralized planned economy and a lower level of democracy – for example, China and Russia – where the State is directly liable for costs and risks of the nuclear facilities and where often there is a strong link between civil and military nuclear interests.

Western nations have so far preferred to have the licensing authorities require that the lifetime of nuclear facilities already in operation be extended through upgrades at costs about a quarter of that of a new facility. After Fukushima, these upgrades are subjected not only to public opinion, but also to a critical scrutiny by experts. In the final analysis, there is a need to include also the economic, social, and political costs – all of which are unpredictable but in any case very high – imposed by the need to dismantle and monitor the radioactive wastes that remain dangerous for virtually an infinite time.

Should Italy Go Back to Nuclear Energy?

After the 2011 referendum, the development of nuclear energy in Italy no longer appears as a viable option. It's worth pointing out that a return of Italy to nuclear power would have been a strategically wrong choice.

The Italian Government had decided to return to nuclear energy in June 2008. The decision led the electrical utility ENEL to enter into a preliminary agreement with EDF (Electricité de France) for the purchase of four 1600-MW EPR reactors that were to be manufactured by Areva. What followed was nothing but a devious and expensive advertising campaign in favor of the nuclear option, this campaign being run by ENEL and the Italian nuclear power Forum, an association created with the contribution of the international nuclear lobby. The campaign was based on arguments that appeared well founded at first sight, but in reality could easily be refuted on the basis of scientific data and on available economic data – even *before* Japan's Fukushima disaster.

Let's examine these arguments one by one:

1) **Nuclear energy is witnessing strong growth in the world.** This statement is simply not true, as we showed in Chapter 6.

2) **Italy needs to find new sources of electric power.** The amount of electric power actually available (123 GW) is already much greater than what is needed (56.5 GW is the peak consumption for a few hours of the year). Hence, nuclear installations would have found strong competition if they had been built. Thus, the request by ENEL to the Government was to give priority to consumption of electricity produced from nuclear power, even if the costs were higher than costs of electricity generated by other sources.

3) **The return to nuclear power is a step toward energy independence.** This is an unfounded assertion, because Italy has neither any uranium mines nor the industrial capability to produce enriched uranium used to power the nuclear reactors.

4) **The use of nuclear energy produces no greenhouse gases.** Even this is not true: building nuclear power plants, feeding them with uranium, freeing them from nuclear wastes and finally decommissioning them would require a considerable consumption of fossil fuels.

5) **Nuclear energy contributes to the revival of Italian industry.** To disprove this claim, 100 Italian industrial managers of the Kyoto Club association have issued a manifesto that underlines how a return to nuclear power would divert the much needed financial resources to the detriment of plans for energy efficiency and renewable energy development, both of which would lead to a significant increase in employment.

6) **Sites for new nuclear facilities are easy to find.** This argument is also misplaced and erroneous. Most of Italy is a seismic area. There is a shortage of water for cooling the reactors. Italy is a densely populated country.

7) **The problem of disposal of radioactive wastes is solved.** In fact it is not, even in the USA, as already discussed. In Italy, there is a waste problem. No region of Italy is even willing to accommodate normal domestic wastes, let alone storage facilities for radioactive wastes.

The return of Italy to nuclear power would be an adventure full of unknowns. Because of the long lead times for issuing permits and the identification of sites (3–5 years), the construction of nuclear power stations (at least 10 years), the operating period needed to amortize the facilities (30–40 years), the dismantling of the facility at the end of its usefulness (50–100 years), and the radioactivity from spent fuel (half-life, hundreds of thousands of years), nuclear energy would be a gamble with the future that is very difficult to assess, not only in terms of social issues, but also economic ones.

In practice, nothing concrete was done between 2008 and 2010 – even the finding of suitable sites to locate the nuclear facilities. Meanwhile, political forces that opposed a return to the nuclear option called for a 2011 referendum.

After the incident of Fukushima, the Italian Government attempted in every possible way, shape and form to defuse the referendum – but despite invitations to abstain, the boycott by the media and deceptive advertising, in June 2011 the necessary *quorum* to validate the referendum was reached. Of the 57% of people who voted, 95% were in favor of abandoning the nuclear option, thus establishing once and for all that Italians did not want Italy to return to nuclear power within its borders.

The Fate of Nuclear Energy

After three consecutive years of decline, electricity generated worldwide by nuclear power installations increased by 2.8% in 2010 compared to the previous year. However, after the Fukushima accident, it is expected that in 2012 there will be a significant decline.

European Union Governments have taken the solemn undertaking to perform checks and verifications (*stress tests*) which, if taken seriously, could lead to the closure of a number of the 146 reactors present in the EU countries. Unfortunately, the tests will be carried out under the responsibility of individual nations. They are not required to disclose the results of the tests.

After Fukushima, Germany took the important step of quickly shutting down one reactor permanently and stopped several others for control inspections. Then the German Government also decided to decommission 8 other reactors and perhaps others later, maintaining either active or on standby only 3 reactors until 2022. In that year, Germany will be the first industrial power to renounce atomic energy as a provider of the 22% of its electricity needs. Switzerland has suspended procedures for the approval of 3 new nuclear facilities and then launched a plan for the gradual decommissioning of its 5 nuclear facilities between 2019 and 2034 that provide 40% of its electricity requirements.

In France, often referred to as a model nation in the growth of nuclear power, it is expected that the share of nuclear over the total installed electric capacity will drop to 44.4% in 2020 and 40.6% in 2030 (was 55.9% in 2006). While its 58 reactors are inescapably aging, only one new reactor is currently under construction, and its entry into the network, initially scheduled for 2013, will most likely suffer a delay of at least three years, with a significant increase in costs. These facts have undermined the credibility of the French nuclear industry, as evidenced in the

Roussely Report of May 2010. In May 2011, the price of shares in Areva dropped by more than half what it was in July 2008.

France's situation is very delicate in that 44 of its 58 reactors are located near rivers, so that during the summer months there may not be enough water flow to ensure the full efficiency of cooling the reactors, as occurred earlier in 2003, 2005, and 2006. It would seem that the situation is reversed in France–it is not the nuclear that helps to combat climate change, but climate change that seems to oppose the growth of nuclear energy. The impossibility of continuing to import electricity from Germany, as France has done every year since 2004, will no doubt aggravate that country's problems.

In June 2012, Japan, whose 30% of electricity came from nuclear installations before Fukushima, has all 50 of its functional nuclear reactors offline. The six Fukushima Daiichi nuclear reactors will have to be dismantled. The large Hamaoka facility, consisting of 3 new reactors intended to be fueled with MOX, is located on a seismic fault and is located only 200 km from Tokyo. This facility will remain offline for at least two more years and perhaps it will then be shut down forever. Other reactors will remain inactive for a long time. The plan to build 14 new reactors over the next 20 years has been abandoned. Japan has decided to develop renewable energies, notably wind power, of which it has plenty.

In the United States, all 104 nuclear reactors, which were constructed before 1980, are presently operating. Twenty-three of these reactors are very similar to those of Fukushima and have raised serious concerns, especially in the case of the nuclear installation located near the Ocean and those in seismic zones. The United States Nuclear Regulatory Commission has issued a dozen new safety rules. There are plans to build two new reactors by 2017, but after Fukushima the percentage of citizens favorable to nuclear energy has dropped from 49% to 41%, and, most important of all, investors are not likely to invest in nuclear power.

China, with its 16 operating reactors and 26 under construction, and South Korea, with 21 operating reactors and 5 under construction, will reinforce controls and safety standards. Apparently, they will continue with their plans to further nuclear development. More uncertain is the situation in India, a country that has not signed the nuclear non-proliferation treaty. There are currently 20 small reactors in operation that cover only 2% of India's electricity consumption (here is a glaring case of civilian nuclear power used as a pretext for military use).

In various other countries–the United Kingdom, Belgium and Sweden–there is a long-standing debate as to whether to continue with nuclear energy or abandon it altogether. Certainly, the disaster of Fukushima will have a much stronger negative impact than Chernobyl on the future of nuclear power in these countries. The Fukushima accident has demonstrated that even a country at the forefront of technology cannot ensure nuclear safety. According to the Japanese Government, it appears that dismantling and cleaning-up the Fukushima site may take some 40 years to achieve, leaving 80 000 people unable to return to their homes.

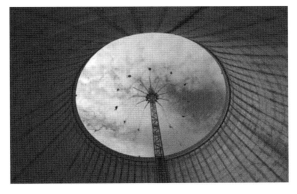

Figure 25 A carrousel inside the cooling tower of the Kalkar nuclear power station in Germany, completed in 1986 but never operated. It is now a popular amusement park with 600 000 visitors every year. It's a pity that radioactivity makes it impossible to convert disused nuclear installations for the same purpose! (Photo: Patrik Stollarz/Staff/ AFP/Getty Images).

In the mean time, problems of every type and gigantic costs continue to block the future development of nuclear fission based on *fast-neutron breeder reactors* (FNBR) mentioned in Chapter 6.

After the costly failures of the European Super-Phoenix and the Japanese Monju undertakings, it is hoped that the fate of the self-breeder reactor facility in Kalkar, Germany, will have some symbolic value. After an outlay of 3.5 billion Euros, the facility never entered into operation. It was converted into an Amusement Park (see Figure 25).

Global Expansion of Nuclear Power?

For various reasons: No! It would be a mistake to expand nuclear power globally. First of all, from a technical point of view, there is a close connection and a strong economic synergy between civilian and military nuclear facilities, as evidenced by the discussions on the development of nuclear power in Iran and North Korea. A general proliferation of civilian nuclear power would inevitably lead to the proliferation of nuclear weapons, and therefore to strong tensions between nations – not to mention the possible increased likelihood of theft of radioactive materials that could be used in devastating attacks by terrorists.

It is also apparent that, because of its high technological content, nuclear energy is increasing the inequalities between nations. Solving the energy problem on a global scale through the expansion of nuclear technology would inevitably lead to a new form of colonization: from the most technologically advanced countries to the least developed nations. Nuclear energy is particularly unsuitable for countries with poor financial, scientific, and cultural resources – even for those countries which have the right to increase their energy availability in the coming years.

Is It Worthwhile to Get Energy Using Technologies Exposed to Great Risks?

The three biggest nuclear accidents were triggered by different causes – (i) technical failure at Three Mile Island, (ii) human error at Chernobyl, and (iii) natural events at Fukushima – all demonstrating that nuclear energy is fragile on all fronts, like any too bold and too complex human enterprise.

It's been said that, with the benefit of hindsight, the Fukushima disaster could have been avoided if Japan had chosen to exploit the country's extensive renewable energy base. Japan has a total of 324 GW of achievable potential in the form of onshore and offshore wind farms (222 GW), geothermal power plants (70 GW), additional hydroelectric capacity (26.5 GW), solar energy (4.8 GW) and agricultural residues (1.1 GW).

We must also recognize that current nuclear facilities are based on an outdated technology, and that the times of the supersonic Concorde – retired from the scene for several years now – and of the *space shuttle* – which accomplished its last space mission only a few months after the disaster of Fukushima – are over.

Climate change caused by the use of fossil fuels, environmental disasters such as that caused by BP's Deepwater Horizon platform in the Gulf of Mexico, the accumulation of plutonium and radioactive wastes, and the repeated incidents of nuclear facilities (of the 594 reactors built since the beginning of the civilian nuclear era, at least 6 reactors have already undergone meltdown) show that we run enormous risks in procuring energy in large quantities and in a very concentrated form.

It is not unlikely that even the extraction of unconventional hydrocarbons, such as *shale gas*, now in place in different countries, the possible exploitation of methane hydrates contained in the oceans, and other techniques to exploit nuclear energy can confront us with precarious problems that no one has yet assessed, let alone addressed.

Fossil fuels and nuclear energy now belong to the past. The future is with energy conservation and better production efficiency of energy based on renewable sources produced from a large number of small and medium size delocalized facilities.

Mankind, however, will be unable to easily shoulder the burden of satisfying its exaggerated voracity for energy.

The bill for our expensive energy lunch, unfortunately, will be left largely to innocent future generations to pay.

10
Energy Italy

In Italy, the shortest line between two points is the arabesque.

Ennio Flaiano

During the brief campaign of the 2011 referendum, Italian television networks aired heated debates on nuclear energy. Unfortunately, one factor that had long dictated a much needed strategy for Italy's energy choices was almost always ignored on those occasions.

Stop Navigating Blind

According to Directive No. 28 of 2009 – better known as *20/20/20* – by 2020 the European Union will have to bring the share of renewable energy on the total *final* energy consumption to 20%, reduce the emissions of CO_2 by 20%, and, through greater efficiency, reduce the consumption of energy by 20%. This may seem a wishful dream, but in fact some European countries are pushing to raise the threshold of renewable energy production even more – to 30%.

In the current legislative framework, every European country has its own specific road map, dictated by historical and economic reasons. On the basis of the principle of *burden sharing*, central governments will have to distribute the total burden evenly among peripheral administrations (i.e., regions, provinces, municipalities). No one can say: *It's not my responsibility* as is often heard in Italian circles.

From 2005 (reference year) to 2020, Italy's share of renewable energy as a fraction of its *overall final consumption* will have to increase from 5.2% to 17% – this to include not only electricity but also energy for the transportation sector and for heating.

Countries that fail to comply with the agreed upon obligations will have to purchase energy from countries that claim a surplus production of energy from renewable sources. Starting from 2020, these countries will then dictate the price to the less virtuous countries. It is expected that they will not be very lenient, financially speaking.

In this binding context, it was expected that the Italian elected officials would act vigorously so as to be prepared for that fateful date of 2020. The exact opposite

Powering Planet Earth: Energy Solutions for the Future, First Edition. Nicola Armaroli, Vincenzo Balzani, and Nick Serpone.
© 2013 Wiley-VCH Verlag GmbH & Co. KGaA. Published 2013 by Wiley-VCH Verlag GmbH & Co. KGaA.

has occurred! On the one hand, Italy has lost time in useless discussions on nuclear power, a technology based on non-renewable sources that in Italy would have begun to produce energy only in (maybe) 20 years, if not later. On the other hand, the 2010 action plan of the Italian Government was to raise the white flag, expecting to import renewable energy for 2020, presumably electricity – easier to transport – equivalent to an energy production quota from a mega electrical power plant of 1500 MW. In a jolt of further self-flagellation, in 2011 the Italian Government reduced the subsidies to renewable sources, creating panic in one of the most exuberant industrial sectors of the country's economy.

This whole affair – which many Italians are unaware of – will within 10 years place Italy at a severe economic and technological disadvantage. Even if Italians were seriously engaged and completely changed course, what could they do? Experts, such as Leonardo Setti of the University of Bologna and the Energy and Strategy Group of the Polytechnic Institute of Milan, have proposed some possible courses of action on this matter.

Conserve Energy! Where? How?

The distribution of total final consumption of energy in Italy is roughly 50% thermal, 30% transportation, and 20% electricity. Insofar as the use of primary sources is concerned, gas dominates the production of heat (65%) and electricity (50%), while the transportation sector is fueled primarily by petroleum products (97%).

In essence, the EU directive has forcefully asked for a 20% decrease of energy consumption by 2020. For the first time in Italy's history, a current undisputed fact has been translated into law. Today's technology allows Italians to live well (even better than before) while consuming less energy than they did 10 years ago. In other words, the 2020 European energy basket must be reduced by one fifth compared to current use. At present, this basket is full of holes.

Italy will have to reduce its overall final energy consumption from 135 Mtoe (in 2005) to 108 Mtoe (in 2020). The saving of 27 Mtoe is equivalent to 310 TWh of electricity or to 33 billion cubic meters of natural gas – people can use whichever unit of measure they prefer.

To achieve this ambitious goal, a series of actions can be taken over the next decade that may lead to a saving of approximately 40 Mtoe. This way, Italy can guarantee that it can achieve, by a good margin, the minimum reduction agreed upon with Europe for 2020 if it does the following:

1) Some 5.5 Mtoe can be saved through a widespread campaign of education on responsible energy consumption through the mass media and schools.

2) Some 1 Mtoe can be saved by eliminating the use of electricity for heating water for use in washing machines, dishwashers and boilers; natural gas should be used instead or, where possible, solar thermal panels (not photovoltaics) used directly.

3) Another 15 Mtoe can be saved by replacing all electrical equipment in current operation (appliances, lighting) with more efficient ones already available commercially, and by improving the efficiency of industrial electrical devices.

4) Another 12 Mtoe can be saved through inspection and certification on the use of energy of at least 70% of buildings, that is, approximately 18 million such structures.

5) And finally some 6.5 Mtoe can be saved by reducing consumption in the transportation sector through measures on vehicles (tires, lubricants), on the behavior of drivers (*eco-driving, car sharing*), on infrastructure (dynamic control of traffic lights, road surface, freight management), and lastly (and why not?), by a serious fiscal disincentive on gas-guzzler cars and by a reduction of highway speed limits to 120 km/h.

With these actions, the energy saved in residential and industrial sectors (excluding transportation) would be largely electricity and heat in amounts equivalent to 27 billion cubic meters of natural gas.

Italy – a Country with an Abundance of Sunlight

For anyone who is seriously involved in energy issues, the disregard of solar thermal energy in Italy can only be a source of unbearable and incredulous embarrassment.

To reach European levels, Italy will have to install by 2020 solar thermal panels whose total area should be at least 25 million square meters. In 2011, Italy had only 2.1 million square meters covered. Hence, it will have to add more than 2 million square meters of new panels every year for the next 10 years, as is being done in Germany. This operation would result in annual savings in 2020 of 2.5 billion cubic meters of natural gas that it is currently being thrown out the window – actually from the rooftop.

Insofar as photovoltaics are concerned, in July 2012 Italy's installations exceeded the threshold of 14.6 GWp power.[1] Its photovoltaic pool, therefore, produces more electricity than would be produced by two 1000 MW power plants. This production, which critics have belittled because it is concentrated under daylight, has – for this very reason – an important economic value. It is available in times of peak consumption, when the demand and the price of electricity are at a maximum.

A minimum objective for 2020 is then to double the photovoltaics pool so as to reach a quota of 32 GWp. These installations would produce 38 TWh/year, approximately 12% of Italy's electricity consumption, and thus lead to a saving of 4.4 Mtoe of natural gas (5.2 billion cubic meters) and 1.4 Mtoe of solid and liquid fuels.

---

1) The Wp (watt-peak) measures the power that a photovoltaic panel produces under standard conditions of operation, with a solar radiation equal to 1000 W/m² at 25 °C. At Italy's latitudes, this peak can be reached within hours during the day; in practice, installing 1 kWp produces on average (every day) 3–4 kWh of electricity.

This amount of electricity, produced locally by millions of citizens and businesses, would amount to half the production of the EPR nuclear reactors, which, according to a Government project – but abandoned as a result of the 2011 referendum – would have had to operate under the strict control of two multinationals. By comparison, those *ghost* nuclear reactors would not have produced even a kWh of useful energy before 2020.

The Italian photovoltaic (PV) effort already produces a quantity of energy that increases day by day. To underline just how modest the objective of 32 GWp for 2020 is, suffice it to note that Germany installed cumulative 24.7 GWp at the end of 2011, and aims to bring this up to 52 GWp by 2020.

An area in which Italy has accumulated delays is that of concentrating solar radiation, which has already been discussed in Chapter 7. As a minimum goal for 2020, Italy should give itself the objective of building two 50-MW installations with heat storage so as to produce about 0.5 TWh of electricity.

Active participation of Italy in the Desertec project (Figure 26), which seeks to share the renewable resources (wind, solar, geothermal, hydro-electricity) of Europe, North Africa, and the Middle East, could significantly increase its share of electricity by concentrating sunlight. In this regard, Italy's enviable geographical location in the heart of the project (see Figure 26) would make it a key partner in the transportation infrastructure.

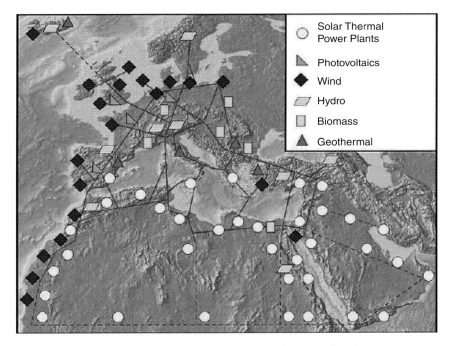

Figure 26 A schematic representation of the energy network expected from the Desertec project/Source: http://en/wikipedia.org/wiki/Desertec.

Wind, Geothermal Energy, Biomass

At the end of 2011, wind power installations in Italy had a cumulative capacity of 6700 MW. A minimal target for 2020 is to increase it to 16 000 MW to produce 27 TWh of electricity, equivalent to 8% of the national electricity needs. This would cut natural gas consumption by about 3 billion cubic meters (i.e., 2.5 Mtoe) and save an additional Mtoe of fossil fuels. Another important goal is to double electricity production by the geothermal option until it reaches 10 TWh/year, thereby saving an additional 600 million cubic meters of natural gas.

Biomass could play a strategic role within the Italian framework – an indirect solar fuel that can be stored and that could therefore be an essential component of an integrated management of renewable energy.

For lack of space, we'll just mention a few figures without going into any detail. The total *sustainable* potential of domestic biomass – *not imported* – is estimated at 15 Mtoe/year: half is agricultural and industrial residues, a quarter is forest biomass, and the remaining quarter is made up of dedicated crops. At present, Italy exploits only about 15% of this potential – it would be desirable to aim at a target of 50% through:

- wherever possible, using wood biomass for domestic and industrial heating through centralized heating systems;

- using biogas generated from agri-food scraps and introduced into the national network of natural gas distribution (this already occurs in Germany – not yet legal in Italy);

- using biogas produced in the framework of an integrated management of municipal solid wastes; and

- using biodiesel and bioethanol produced from dedicated crops.

The implementation of renewable energy production facilities requires three essential elements: scientific rigor, planning, and the active involvement of the local citizenry. Recent years have often witnessed the realization of projects that were very deficient in all these respects. Not infrequently, unfortunately, criminal organizations have penetrated this sector, always ready to exploit any type of highly profitable activities.

Creating new installations for the production of renewable energy is essential. Nevertheless, citizen groups that are accused of suffering from the *not in my backyard (NIMBY) syndrome*, because they oppose such installations as wind farms, sometimes are right.

Unfortunately, Italy is studded with poorly conceived or poorly managed projects by rapacious and incompetent entrepreneurs who too often exploit situations generated by ambiguous regulations that are incompatible with serious energy planning. In these situations, local authorities are often caught between the economic interests of the proposers and the protests of their citizenry, who have already been scarred by previous experiences.

If Italy fails to put an end to poor planning and mismanagement, the objectives of the EU 2020 project will remain a mirage, missing out on a unique opportunity for Italy's industrial, economic, and environmental revival. The credibility of the projects is the key to Italy's energy transition. Throwing this out the window would be nothing less than irresponsible.

Conservation and Renewables – a Summary

In previous pages we have listed some goals of energy saving and renewable energy production that Italy needs to achieve in the next decade. The concept is very simple: by 2020 Italy should save a large amount of natural gas, quantifiable into 45 billion cubic meters equal to 50% of its current consumption. This saving could be achieved through two strategies: (i) more efficient consumption of heat and electricity and replacement (as far as possible) of fossil fuels used for electricity production, and (ii) exploitation of the whole range of currently available technologies that make use of free and perpetual primary sources – the Sun, wind, water, and the endogenous heat of the Earth.

What can Italy do with the small treasure of natural gas saved by the abovementioned actions? It could use it as a source of energy for transportation, pushing the methanization of the large pool of Italian vehicles presently on the road to the maximum. In other words, Italy should rely on efficiency and renewable resources to decrease its dependence from oil. Once Italy achieves such ambitious goals, renewable energy sources will likely have been established so that this country can safely schedule the end of its consumption of fossil fuels by the end of the twenty-first century.

This historic operation is technically possible. Does Italy have a governing class that can take up the challenge? What happened in recent years seems to prove otherwise, but only time will tell!

11
Energy Canada

A lot of people like snow. I find it to be an unnecessary freezing of water.

Carl Reiner

No one would question that advances in technology and the discovery of new sources of energy over the last couple of centuries have led to significant economic development, which took place mostly in Europe and North America. As a case in point, the once-flourishing iron-making industry in England had to move twice as a result of shortages of wood (source of charcoal), first to Ireland and then to Scotland to be closer to abundant supplies of wood. Charcoal was then replaced by coal in the nineteenth century. In turn, in many of its uses coal was replaced by oil (petroleum) a century later. For various reasons – not least those environmentally related – coal may soon be displaced by natural gas.

In the long run, renewable energy will displace fossil fuels, which have so far left a large carbon footprint when combusted. Deleterious effects of this carbon footprint on planet Earth cannot be sustained for much longer. Mankind will have to embrace non-polluting renewable energies to minimize these effects, if not to suppress them altogether. However, this will not occur without significant policy shifts by Governments through subsidies. When large quantities of renewable electricity become available, maybe part of it will be converted into hydrogen (hydrogen-based economy).

Most of the Earth's surface is water. Unfortunately, at current costs, a hydrogen-based technology requires more energy to produce the hydrogen (an energy vector) from water than it can provide – in essence the payback is negative. Others maintain that nuclear fusion may eventually become the major energy contributor to satisfy the world's energy demand. At present, the nuclear fusion technology is still in its infancy. Accordingly, fossil fuels will continue to be – for some time to come – the major sources of energy.

Primary Energy Resources

Canada is a resource-rich country with vast reserves of coal, oil, natural gas and uranium, not to mention the many rivers that have yet to be exploited for

Powering Planet Earth: Energy Solutions for the Future, First Edition. Nicola Armaroli, Vincenzo Balzani, and Nick Serpone.
© 2013 Wiley-VCH Verlag GmbH & Co. KGaA. Published 2013 by Wiley-VCH Verlag GmbH & Co. KGaA.

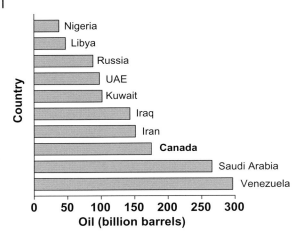

Figure 27 World's proven oil reserves. Source: *BP Statistical Review*, 2012.

hydroelectricity. In such a large country, spanning nearly 7000 kilometers from coast to coast, many areas can also be exploited for wind, solar, and tidal power projects.

Presently, Canada uses fossil fuels as the source of energy to meet almost all of its transportation needs, and uses hydroelectric dams, fossil fuels, and nuclear power plants to satisfy most of its electrical needs. All these energy sources provide Canada with a good, flexible energy mix, which can be used to offset increases in energy costs such as those experienced by the fossil fuel industry in the not too distant past. Having an energy mix also gives Canada some energy security.

Following the oil crises of the 1970s, many scientists and oil industry specialists warned that we would soon run out of fossil fuels. However, recent studies now indicate that natural gas and oil may well last into the twenty-first century. According to Natural Resources Canada (NRCAN), Canada holds 8.7 billion tons of proven coal reserves, including 6.6 billion tons of proven recoverable coal reserves, which may be sufficient for a hundred years at current production rates of consumption. But whether it is wise to continue using fossil fuels for so long is, of course, another story.

Figure 27 reports the estimated world's proven oil reserves.[1] A recent estimate by NRCAN indicates that Canada also has significant proven reserves of crude oil, after those of Saudi Arabia and Venezuela, although these are mostly from the oil sands in the Province of Alberta. Canada's current total oil reserves are estimated at about 180 billion barrels including the oil sands, which Government regulators estimate to be economically recoverable at current costs and with current technology.

Though Canada's present oil reserves are huge, the Chief Executive Officer of Shell Canada and other experts optimistically estimate that the amount of oil that can be recovered from oil sands may be closer to 2 trillion barrels or more – this would make Canada's reserves 8 times greater than those of Saudi Arabia.

---

1) All reserve estimates involve a great degree of uncertainty, depending as they do on the amount of reliable geological and engineering data available and the proper interpretation of those data.

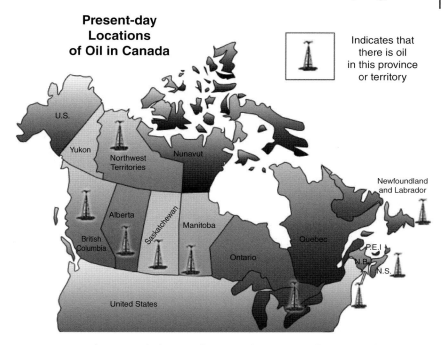

Figure 28 Current locations of oil in Canada. Source: http://www.collectionscanada.gc.ca.

Over 95% of Canada's reserves are in the oil sands deposits in the province of Alberta; nearly all of Canada's oil sands and much of its conventional oil reserves are located here. The balance is concentrated in several other provinces and territories. For instance, Saskatchewan and offshore areas of Newfoundland, in particular, have substantial oil production and reserves: Alberta, 39% of Canada's remaining conventional oil reserves, offshore Newfoundland 28%, and Saskatchewan 27%. If oil sands were included, Alberta's share would be over 98%.

Alberta and Newfoundland are not the only places in Canada with oil reserves (see Figure 28). Canada's northern frontier also has oil which is being explored, and in fact, as can be seen from the figure, nearly every province and territory of Canada possesses some oil. A vastly ignored and untapped region is the Old Ontario Oil Belt, where North America's first commercial oil well was drilled in 1858.

Alberta's oil sands occupy 140 200 square kilometers (54 132 square miles) in the Athabasca, Cold Lake, and Peace River areas of northern Alberta. Approximately 602 km^2 of land has been exploited for oil sands mining activity – that is, less than 0.25% of Alberta's Boreal Forest, which covers over 381 000 square kilometers. The oil sands deposits span a region larger than several states in the United States – a region even larger than England.

Canada is not the only country with oil sands reserves, however – several other countries, including Venezuela, the United States, and Russia have similar deposits, albeit not in the same quantity. Alberta's Athabasca deposit is currently the largest and most developed, and the most technologically advanced production processes are being employed here.

Over 99% of Canadian oil exports are sent to the United States through various pipelines–Canada is the United States' largest supplier of crude oil. In 2009, Alberta exported daily about 1.4 million barrels, supplying the United States with 15% of oil imports, or 7% of its oil demand. Total oil consumption by the United States in 2011 was 18.8 million barrels per day (bbl/d). Canada as a whole exported 1.9 million barrels of crude oil daily to the USA, or about 22% of its imports.

Construction of the Keystone Pipeline, that was to transport Canada's oil sands crude to Texas was placed in limbo by President Obama following environmentalists' concerns about the projected route through some ecologically fragile and sensitive areas of Nebraska.

There is much discussion in the United States Congress on this issue as the project would have opened thousands of jobs, though the exact number seems to be a subject of considerable debate. Nonetheless, the project will likely be revisited in the near future.

Since oil sands–sometime also referred to as tar sands–have been viewed with so much suspicion by environmentalists, it's worth examining more closely what oil sands are, together with related environmental concerns.

Oil Sands or Tar Sands?

Oil sands have often been referred to, albeit incorrectly as tar sands. Although visually they appear to be similar, tar and oil sands are quite different. Oil sands are a naturally occurring petrochemical source, whereas tar is a substance produced from the degradation of hydrocarbons. In addition, their uses are totally different–oil sands are first refined to produce oil, and this, after subsequent treatment at an oil refinery, is converted into a useful fuel. Tar, on the other hand, cannot be refined and historically has been used as a wood sealer and for treating rope against moisture. In any case, the two terms are often used interchangeably. Canada's First Nations used oil sands for water-proofing. The earliest reported mining of oil sands occurred in 1745 in north-eastern France, with refining capacity being introduced nearly a century later in 1857.

Alberta's oil reserves play an important role in Canada's and in the global economy, as Canada is a politically stable and reliable supplier of energy to the world. Time Magazine described Alberta's oil sands as *Canada's greatest buried energy treasure*. But what are oil sands exactly?

Oil sands essentially consist of a naturally occurring mixture of sand, clay and/or other minerals, water, and bitumen–a heavy and extremely viscous oil that must be treated before it can be used by refineries to produce usable fuels such as gasoline and diesel.

The Athabasca Oil Sands region of Alberta is vast (see above and Figure 29) and contains a gigantic reserve of oil which has become economically viable

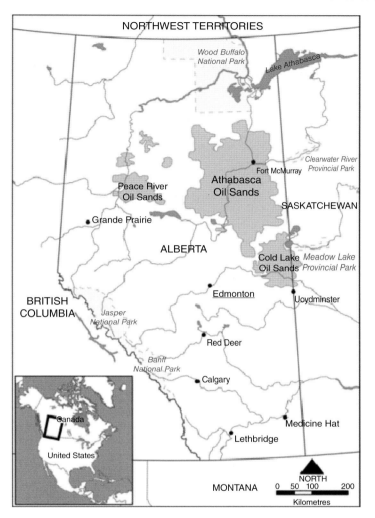

Figure 29 Map of the Province of Alberta (inset: location of Alberta in Canada) illustrating the Athabasca oil sands (grey-colored area) being exploited in Canada.

to mine. However, because global warming is happening at a rate much faster than anticipated, the question is whether it is reasonable to exploit them so intensively.

The most recent estimates of Alberta's oil sands indicate that about 173 billion barrels (27.5 billion cubic meters) of crude oil could be recovered economically – equivalent to about 10% of the estimated 1700 billion barrels of bitumen-in-place. Estimates also showed that there is a potential of about 315 billion barrels of oil that could be recovered in the oil sands provided that more favorable economic conditions are in place and a new technology available for extraction and processing. Germane to this, oil companies that use the steam-assisted gravity drainage technology (SAGD) to extract bitumen have indicated that they could recover over 60% with little effort.

So far, approximately 3% of the initial estimate of crude bitumen reserves have been extracted since commercial production began in 1967. At the projected rate of production for 2015–about 3 million barrels per day (480 000 cubic meters daily)–the Athabasca oil sands reserves alone could last over 170 years. The need for workers in mining the oil sands has driven the unemployment rates in Alberta and adjacent British Columbia to the lowest levels in history.

Canada's Government has recently approved the construction of the Joslyn North Mine in Alberta, giving the go-ahead after a 6-year wait–it is expected that this new mine will inject $9 billion into new capital investment across the country and bring in some $10 billion in revenue to Canada and to Alberta. Clearly, money and jobs are the name of the game.

Oil Sands and Their Environmental Impact

Approximately 20% of Alberta's oil sands are recovered through open-pit mining, with the remaining 80%–because of depth–requiring *in situ* extraction methods.

Canada's Boreal Forest comprises about one third of the circumpolar boreal forest that rings the Northern Hemisphere, mostly north of the 50th parallel. The Canadian boreal region represents a tract of land over 1000 kilometers wide separating the tundra in the north from the temperate rain forest and deciduous woodlands that predominate in the most southerly and westerly parts of the country. The boreal region is home to about 14% of Canada's population. With its sheer vastness and integrity, the boreal forest makes an important contribution to the rural and aboriginal economies of Canada, primarily through resource industries, recreation, hunting, fishing, and eco-tourism.

Needless to say, such open pit mining is an eyesore in the landscape and is destroying the Boreal Forest and muskeg (an acidic soil type common in Arctic and boreal areas, although it is also found in other northern climates).[2] The Alberta government requires that mining companies restore the mined land to *equivalent land capability*–meaning the ability of the land to support various land uses after reclamation similar to what existed previously–although the individual land uses may not necessarily be the same. For instance, oil sands companies are making good progress at reclaiming mined land, but its use is to be as pasture for wood bison (a species related to the American buffalo)–the land is not restored to the original boreal forest and muskeg.

2) Muskeg is nearly synonymous with bog land – however, muskeg is the standard term used in Western Canada and Alaska, whereas bog is commonly used elsewhere. See: http://en.wikipedia.org/wiki/Muskeg.

Water Usage

It takes an awful lot of water to extract one cubic meter of synthetic crude oil from oil sands, about 2–4.5 cubic meters. Currently, mining of oil sands diverts some 360 million cubic meters of water from the Athabasca River – that is, more than twice the amount of water required to meet the annual municipal needs of a city the size of Calgary (the oil capital of Canada located in Alberta). Depending on how much of the water is recycled, to produce one cubic meter of oil with *in situ* production methods could be as little as 0.2 cubic meters. The Athabasca River runs a distance of some 1231 kilometers from the Athabasca Glacier to Lake Athabasca; its average flow is 633 cubic meters per second and reaches as much as 1200 cubic meters per second at its highest daily average.

Natural Gas Usage

The standard extraction process also requires large quantities of natural gas. At present, the oil sands industry uses about 4% of the Western Canada Sedimentary Basin natural gas – this is likely to increase to about 10% by 2015.

According to Canada's National Energy Board (CNEB), production of one barrel of bitumen (heavy viscous oil) from *in situ* projects requires 34 cubic meters (1200 cubic feet) of natural gas, and about 20 cubic meters (700 cubic feet) for integrated projects. This represents a large gain in energy if we consider that one barrel of oil equivalent is about 170 cubic meters (6000 cubic feet) of gas. Were the natural gas supply in the mining operations to run low, Alberta might then be forced to reduce its natural gas exports to the United States. Alternatively, if natural gas reserves were to dwindle away altogether, oil processors would then have to gasify the bitumen to generate their own fuel. In other words, just as bitumen is converted to synthetic crude oil (SCO), it can also be converted to synthetic natural gas (SNG).

In 2007, oil sands operations used about 28 million cubic meters (1 billion cubic feet) of natural gas daily – approximately 40% of Alberta's total usage. The Canadian Energy Resource Institute (CERI) has estimated natural gas requirements in mining operations at 2.14 GJ (2040 cubic feet) per barrel for cyclic steam stimulation projects, 1.08 GJ (1030 cubic feet) per barrel for the steam-assisted gravity drainage (SAGD) technology – an enhanced oil recovery technology for producing heavy crude oil and bitumen – 0.55 GJ (520 cubic feet) per barrel for bitumen extraction in mining operations if upgrading is not included, or else 1.54 GJ (1470 cubic feet) per barrel for extraction and upgrading.

Greenhouse Gases

In spite of the silver lining noted above, the predicted growth of synthetic oil production in Canada has threatened its international commitments with respect to the Kyoto Protocol it signed in 1997. In doing so, Canada had agreed to reduce, by 2012, its greenhouse gas emissions by 6% with respect to 1990 levels. Instead, in 2002 Canada's total greenhouse gas emissions had increased by 24% with respect to 1990. However, oil sands mining contributed only 3.4% to Canada's greenhouse gas emissions in 2003.

Canada has a population of around 33 million (2011 Census), and is the world's eighth largest emitter of greenhouse gases. To remedy the situation, the Canadian Integrated CO_2 Network (ICO2N) initiative intends to promote development of large scale capture, transport, and storage of carbon dioxide (CO_2) as a means of helping Canada to meet its climate objectives while supporting economic growth. ICO2N members include a group of industry participants and many oil sands (major oil) companies.

In spite of this initiative, however– *Emissions in 2010 top 10 billion tons; Canada one of highest emitters*–was the headline in an article signed by Margaret Monroe of the PostMedia News organization. This figure represents a global increase of *carbon* emissions of 5.9% relative to 2003 and nearly 49% based on 1990 figures, even though Canada and other countries had pledged to cut emissions as part of the Kyoto Protocol. Canada's share in 2009 was 690 million tonnes of CO_2, making it one of the largest per capita emitter, though the financial crisis of 2008 and 2009 led to short-lived decreases in global emissions, which were indeed short-lived as they rapidly jumped in 2010. China, the United States, India, the Russian Federation, and Europe were the largest contributors to global emissions growth in 2010 and 2011. Canada's present emissions (November 2011) run 17% higher than 1990 levels, in part due to increased emissions from Alberta's oil sands, which is totally at odds with its initial Kyoto protocol commitments.

Mike De Souza's article–also of PostMedia News–on the United Nations Conference on Climate held in Durban, South Africa (December 2011) quoted a former Canadian government negotiator (Vic Buxton), who worked on one of the world's most successful environmental treaties, on Canada's stance at the 2011 conference: *I'm not sure the objective is to achieve progress– I think the objective appears to me to be to make sure nothing is put in place in an international regulatory sense that can impede economic development in the Alberta tar sands.* Canada's Environment Minister, Peter Kent, was also quoted as saying that *Kyoto is the past*–Canada has walked away from Kyoto so as not to be charged the $16 billion penalty for not adhering to its commitment under the protocol.

Canada is seeking a new international deal to replace Kyoto, a deal with a firm commitment to limit the temperature rise to less than 2 °C. However, unless China and India sign on–the two biggest greenhouse gas emitters–the new deal is a non-starter.

Research was carried out at the University of Victoria (British Columbia) and published in the journal *Nature Climate Change* by one of Canada's most respected climate scientists, Dr. Andrew Weaver, who together with Neil Swart investigated what the impact would be on climate change of producing the 180 billion or so barrels of Alberta's oil-sands crude. The reputed "filthiness" of the oil sands has sparked tough and heated debates around the world, particularly by environmentally conscious people in the United States and in the European Union, so much so that the projected Keystone pipeline that was to bring the crude oil from the oil sands of Alberta to the Texas refineries has been placed on the back burner. The Canadian Government is intent in selling its crude from the oil sands to other

Pacific-rim countries—China with its insatiable thirst for energy has shown some interest. This would require building a new pipeline through the Rockies to the West Coast at a cost likely higher than the projected Keystone pipeline.

Apparently, even if all the oil sands were extracted, a course of action that could take well over 100 years to achieve at the estimated 2015 rate of 3 million barrels a day, Dr. Weaver's analysis shows that the cumulative extent of global warming resulting from the oil sands would be no more than about 0.02–0.05 °C increase in temperature. In addition, even if every barrel of the oil sands was produced from the estimated reserve of some 2 trillion barrels it would only raise the global temperature by about 0.3 °C. By contrast, burning the world's huge coal resources would likely raise the temperature by 15 °C; consuming shale gas would increase the temperature by less than 3 °C. However, not everyone is buying into this analysis, as the complete life-cycle of the oil from the oil sands was not taken into account.

Coal in Canada

In coal reserves, Canada ranked 10th in the world at the end of 2011 with 6582 million tonnes (Table 7c, page 50), and in 2011 Canada produced about 68 million tonnes—of these, about 42 million tonnes were consumed nationally that year.

Historically, coal has been an important resource of Canada—it was used to heat homes during the severe winter months, to power trains across the length of Canada (about 7000 kilometers coast to coast), and to help fuel Canada's industrial development. The coming of advanced technologies has positioned coal as a high-value energy source. In spite of its big carbon footprint (Figure 30), coal remains the largest single source of electricity production worldwide. In addition, coal helps in the production of over 70% of the world's steel, and is used in other industrial processes—for example, in the manufacturing of cement.

Coal will likely continue to play a role in the energy mix of many countries in power generation. In that case, ways to address environmental concerns on at least three fronts:

1) Minimize local air, water, land, and community impacts at the mine sites.
2) Improve coal-burning technologies to reduce emissions at the power plants, and use advanced technologies for even greater reductions.
3) Cooperate on a global scale to minimize coal's environmental impact.

On all these fronts, local authorities, communities and other stakeholders must be engaged in addressing important environmental initiatives.

In its recent sobering report on energy demand for fossil fuels and the growth of CO_2 emissions, the energy giant Exxon Mobil forecasts that CO_2 emissions from energy sources will see an increase until 2030—they will then level off as natural gas replaces coal. Exxon's annual outlook to 2040 also indicates that total energy demand will increase by 30% over the next three decades—oil, natural gas, and coal contributing up to 80% of supply. More than 90% of the additional energy demand will come from China and other developing nations. The report also

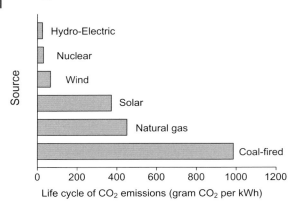

Figure 30 Graph illustrating the contribution of coal – specifically – and other fuels used in electricity generation to the life cycle of CO_2 emissions. Data source: http://www.cn.ca.

points out that CO_2 emissions are on the decline in North America, Europe, and other developing economies – China will apparently follow suit only around 2025, ending decades of rapid economic development.

Natural Gas

Natural gas consists primarily of methane, and some quantities of ethane, propane, butane, pentanes, and heavier hydrocarbons. It formed naturally from carbon and hydrogen atoms from organic matter after millions of years – it is found trapped within Canada's geological formations. Natural gas is a reliable, secure, safe, and environmentally acceptable fuel. According to the Canadian Association of Petroleum Producers (CAPP), natural gas is an abundant natural resource of Canada – the world's third largest producer of natural gas, with an average annual production of about 5.9 trillion cubic feet (tcf); see also Table 6B. At today's consumption levels, North America can count on a natural gas supply for a century.

Canada's provinces of British Columbia, Alberta, Quebec, Nova Scotia, and the Northwest Territories all have significant natural gas resources. Exploration of natural gas reserves offshore is ongoing in Nova Scotia, along with shale gas in northeastern British Columbia and Quebec. However, the majority of commercial production of natural gas is currently taking place in Western Canada. With other energy sources, natural gas is the cornerstone of Canada's future energy supply mix.

CAPP has advocated diversification of the energy supply mix and the use of the right fuel in the right place at the right time. Natural gas is expected to play an important role in this scenario.

The wide energy mix available to Canada provides it with some strength. For example, nuclear, fossil fuel, and hydroelectricity are key for base-load electricity generation. As solar and wind power technologies become more fully developed, they should contribute significantly in this capacity.

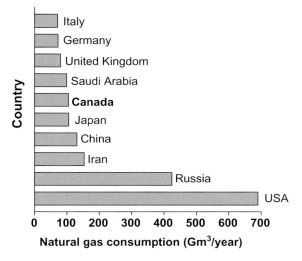

Figure 31 Graph showing the top 10 nations in the global consumption of natural gas in 2010 (billions of cubic meters per year). Source: *BP Statistics World Energy Review*, 2012.

Global natural gas consumption (in billions of cubic meters) is illustrated in Figure 31. In 2011, Canada ranked sixth as a consumer of natural gas, yet it only has a population of about 33 million people.

Natural gas can be obtained from **unconventional** and **conventional** sources, the major difference between the two being the manner, ease, and cost of extracting the resource. An assessment of global proven reserves of natural gas by the end of 2011 carried out by British Petroleum (*BP Statistical Review*, 2012) and given by regions together with the European Union, the United States, Canada, the United Kingdom, and Italy is summarized in Table 10 – Canada's share of the world's reserves is about 2.0 trillion cubic meters or 1.0% of total.

Conventional natural gas is typically *free gas* trapped in multiple, relatively small, porous zones in naturally occurring rock formations (carbonates, sandstones, and siltstones). However, most of the growth in supply occurs from recoverable gas resources found in unconventional gas reservoirs that include "tight" gas (an unconventional natural gas difficult to access because of the nature of the rock and sand surrounding the deposit), coal bed methane, gas hydrates, and shale gas. Technological breakthroughs in horizontal drilling and fracturing have made shale and other unconventional gas supplies commercially viable, revolutionizing Canada's natural gas supply picture (Figure 32).

Nuclear Energy and Electricity

Nuclear energy accounted for nearly 13.5% of the world's electricity in 2011. Nuclear became an important energy source after the first oil crisis of autumn 1973 because of the low cost of nuclear fuel compared to other primary energy

Table 10 World estimates of *proven* natural gas reserves in trillion cubic meters by regions; reserves for the European Union, the United States, Canada, the UK, and Italy are also shown for comparison.

Regions/Countries	Reserves (trillion m³)	% Share of Reserves	Reserves/Production Ratio
North America	10.8	5.2	12.5
South and Central America	7.6	3.6	45.2
Europe and Eurasia	78.7	37.8	75.9
Middle East	80.0	38.4	–
Africa	14.5	7.0	71.7
Asia Pacific	16.8	8.0	35.0
World Total	**208.4**	**100**	**63.6**
European Union	1.8	0.9	11.8
United States	8.5	4.1	13.0
Canada	2.0	1.0	12.4
UK	0.2	0.1	4.5
Italy	0.1	≪0.05	11.4

Source: *BP Statistical Review*, 2012.

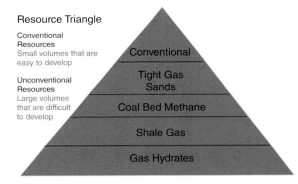

Resource Triangle

Conventional Resources
Small volumes that are easy to develop

Unconventional Resources
Large volumes that are difficult to develop

Conventional

Tight Gas Sands

Coal Bed Methane

Shale Gas

Gas Hydrates

Figure 32 Conventional and unconventional resources of natural gas. Source: http://www.capp.ca/canadaIndustry/naturalGas/Conventional-Unconventional/Pages/default.aspx#qow5TnMP83A9.

sources and because of abundant uranium reserves (about 5.4 million tonnes) located in large part in politically stable countries. Total known recoverable uranium resources plus historical production levels are reported in Figure 33 – Australia, Kazakhstan, and Canada account for nearly half the known reserves. Although Canada was the largest uranium producer (ca. 22% of world's total) in 2009 it was overtaken by Kazakhstan. Nonetheless, Canada retains the world's largest reserves of high-grade, low-cost uranium.

The country that produces most of its electricity from nuclear power is France (78%), followed by Slovakia and Belgium (54%). Another 14 countries rely on nuclear energy for about a quarter of their electricity supply. The United States has the largest nuclear generating capacity, with 104 reactors that generate about

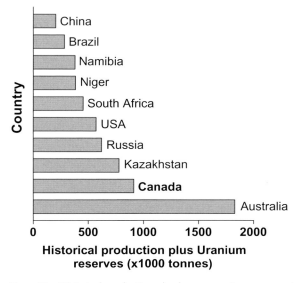

Figure 33 Historical production plus known uranium reserves in thousand tonnes of the top 10 countries. Data source: OECD Uranium 2009: Resources, Production and Demand. OECD-NEA Publication 6891. 2010.

19% of its power needs. By comparison, Canada gets around 15% of its electricity from nuclear power–in particular, Ontario has 16 CANDU nuclear reactors that produce 51% of its electricity. Globally, there are 12 CANDU reactors in six different countries outside Canada–more are in the development phase. Domestically Canada is home to 22 CANDU reactors spread across three provinces.

Canada has developed its own nuclear technology that is quite different from others. The CANDU reactors use fuel pellets of naturally occurring uranium mined in Canada. These pellets are inserted into 0.5 meter length tubes known as fuel bundles that provide enough electricity to power 100 homes for a year. A single 1.65 cm nuclear fuel pellet can produce the same amount of energy as 807 kilograms of coal, 677 liters of oil, or 476 cubic meters of natural gas.

But for major incidents in the last few decades, nuclear power might have grown exponentially. However, Chernobyl, Three Mile Island, and recently Fukushima have left a significant negative impact on the future use of nuclear energy for electricity generation (see Chapter 8).

A fuel bundle

Just like the banking system, which is the most regulated in the world, nuclear power is also highly regulated in Canada. Members of the Canadian Nuclear Safety Commission (CNSC) use the *Nuclear Safety and Control Act* to decide on policies concerning nuclear energy, materials, and equipment. Regulations enforced by CNSC are meant to protect the safety, security, and health of Canadians – they apply to all nuclear activities (mining, refining, spent fuel, etc.) including hospitals and clinics that use radioisotopes in various medical procedures.

Although nuclear reactors burn no fossil fuels and produce no smog or greenhouse gas emissions (GHGs) during their operation, *they do produce* nuclear wastes that need to be handled carefully and stored safely. Canada's long-term strategy for managing spent nuclear fuel is to contain it in a central, isolated facility built in a deep rock formation. In the meantime, new and innovative solutions on how to manage nuclear waste over time are being explored.

Electricity

At present, in addition to its vast reserves of oil, gas, coal, and uranium, Canada is also blessed with major rivers, many of which have yet to be exploited for hydroelectric projects. On the down side, hydroelectric dams often result in flooding large landmasses at some ecological cost. Burning fossil fuels for transportation and electricity generation emits millions of tonnes of greenhouse gases, including sulfur dioxide (SO_2) and nitrogen oxides (NO_x) that contribute to smog and acid rain, whereas nuclear power plants produce highly radioactive wastes. Alternative renewables such as wind, solar, tides, and biomass can only meet a fraction of Canada's electricity needs.

The question being asked in Canadian circles is what role each of these renewables will play in the electricity mix that would best suit Canada's future needs, at the same time with an eye to promotion of clean air and reduction of greenhouse gases. In this regard, Ontario uses special technologies in its coal-fired generation stations that can reduce or almost eliminate some undesirable atmospheric pollutants.

Alberta's electricity production is the most carbon-intensive, emitting around 56 million tons of CO_2 and accounting for 47% of all Canadian emissions in the electricity and heat generation sectors – followed by Ontario (27 million tons), Saskatchewan (15 million tons) and Nova Scotia (9 million tons). The province of Quebec has the smallest carbon footprint in the electricity sector with 2.45 g of CO_2 per kilowatt-hour of electricity generated, because most of it is produced from hydro.

By 2008, Ontario reduced its greenhouse gas emissions significantly as a result of gradual decommissioning of coal-fired electricity generation stations, to be completed by 2014. As a further effort, Ontario commissioned a new natural gas-fired power plant to generate 4700 MW and wind farms to generate 1100 MW of electricity, which allowed it to retire two additional coal-fired units by the end of 2010, on track for a complete phase-out at the end of 2014.

For its part, Alberta has embarked on the construction of a carbon capture and storage facility at the 450-MW Keephills-3 supercritical sub-bituminous coal-fired

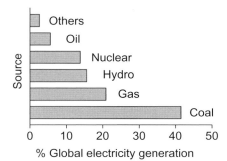

Figure 34 Global electricity generation from various sources.

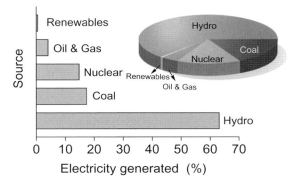

Figure 35 Canada's sources of electricity generation.

power station, which involves enhanced oil recovery and storing the captured CO_2 through geological sequestration – it is scheduled to be operational by 2015. In March 2010, the Saskatchewan Power Utility, SaskPower, announced its own carbon sequestration project at its largest coal-fired plant. British Columbia has decommissioned its 50-year old gas-fired Burrard Generating Station.

Figure 34 illustrates the various sources used in global electricity generation; Figure 35 displays the energy sources Canada uses for electricity generation.

Canada was the world's third-largest producer of hydroelectricity (after China and Brazil) in 2011, accounting for about 63% (372.8 TWh) of all its electricity production – other sources are coal (16.5%), nuclear (15.2%), oil and natural gas (4.1%), and renewables (1.3%). By contrast, 41.5% of electricity generation globally comes from coal, 20.9% from natural gas, 15.6% from hydro, 13.8% from nuclear power, 5.6% from oil, and 2.6% from other sources.

Large hydroelectric projects have been built since the early 1960s, especially in Quebec, British Columbia, Manitoba, and Newfoundland. This has led to significant increases in Canada's generation capacity. By contrast, the Canadian-designed CANDU nuclear reactors supplied more than half of Ontario's electricity demand in 2007.

Canada generated 617.5 terawatt-hours (TWh) of electrical power in 2007, placing it seventh worldwide. There are about 822 generating stations scattered from the Atlantic to the Pacific with a total capacity of 124 240 megawatts (MW) – compare this with 111 000 MW in 2000. A hundred large generating stations have a combined capacity of 102 341 MW. In 2009, Canada's electricity production was 18 566 kWh per capita with domestic use accounting for about 94% of production (17 507 kWh per capita). Much of Canada's domestic electricity use (64.5%) was produced with renewable sources (rivers, etc.). Compare this with OECD countries, where the average electricity production per capita was 8991 kWh in 2008. The electricity use from non-renewable sources (fossil fuels, nuclear energy) in Canada was 6213 kWh per capita in 2009 – it was 5579 kWh in the United Kingdom, 5811 kWh in Germany, 4693 kWh in Denmark, 4553 kWh in Spain, 11 495 kWh in Finland and 12 234 kWh in the United States.

Renewable Energy – Wind Power

Using wind energy reduces the environmental impact of generating electricity because it requires no fuel and produces neither pollution nor greenhouse gases. Wind power has had a long history in Canada that dates back several decades, especially in remote farms of the Prairie Provinces (Manitoba, Saskatchewan, and Alberta) that used this renewable source to pump water and generate electricity. However, the amount of electricity generated by wind power in Canada remains small relative to hydro and coal sources – nonetheless, it is the fastest growing source. As of October 2011, wind power capacity represented about 2% of Canada's electricity demand, with about 4708 MW of generating capacity. In its *Wind Vision 2025* Report, the Canadian Wind Energy Association (CanWEA) outlined a future strategy for wind energy in Canada that should reach a capacity of 55 000 MW by 2025 or about 20% of Canada's energy needs.

In the Canadian context, early development of wind farms took place in Ontario, Quebec, and Alberta, with the latter province having built the first commercial wind farm nearly two decades ago in 1993 – British Columbia added wind power to its power grid in November 2009. With an ever-increasing population, Canada is diversifying its energy supplies away from traditional reliance on fossil fuels and heavy reliance on hydroelectricity, at least in some provinces such as Nova Scotia, where as much as 12% of its electricity comes from renewable sources. An additional 2004 MW of wind power is to come on stream in Quebec by 2015.

A survey conducted in October 2007 showed that 89% of Canadians favored renewable energy sources (wind and solar power) as these sources were better for the environment. An earlier survey (April 2007) indicated wind power as the alternative energy source most likely to gain public support for future development in Canada – 16% opposed this type of energy source. Unfortunately, just as in Italy there is local opposition from residents near wind farms who also suffer from the not-in-my-back-yard (NIMBY) syndrome because of perceived eyesore, noise, and loss of property value. Yet, wind farms in some parts of Canada have become tourist attractions, much to the surprise of the owners of the wind farms. Three out of four Canadians oppose nuclear power development.

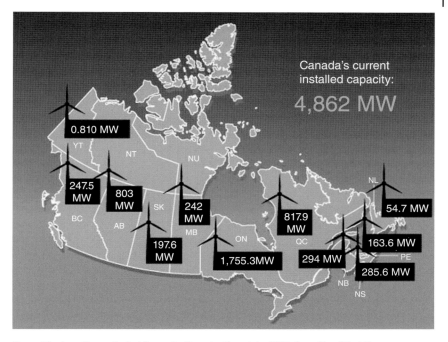

Figure 36 Locations of wind farms in Canada; Copyright 2008 Canadian Wind Energy Association (CanWEA). See http://www.canwea.ca/farms/index_e.php.

To overcome this NIMBY syndrome, offshore wind farms prove to be less obtrusive than onshore wind farms as the apparent size and noise of the turbines are mitigated by distance – moreover, the average wind speed is considerably higher at sea than on land. Ontario is pursuing several proposed locations in the Great Lakes region, while other Canadian projects include one on the Pacific west coast. Some of the locations of current on-shore wind farms in Canada are illustrated in Figure 36.

We should distinguish between a *large wind* farm and a *small wind* farm. The latter involves either a small turbine powering a house or a medium-sized turbine powering a farm, a business or a small community. By contrast, a large wind farm provides electricity to the electric power grid. At present, Canada's wind farms have a capacity of 4862 MW – enough to power over one million homes – about 2% of Canada's total electricity demand. Wind power is widely distributed in rural areas throughout Canada with various wind farms in operation – more are under construction.

In summary, Canada's massive hydroelectric resource, that provides 60% of Canada's electricity, complements well with wind energy, that could provide good opportunities to integrate more wind energy into the system than is the case in many other countries.

Renewable Energy – Solar Power

Canada is also blessed with plentiful solar energy resources located in large part in Ontario, Quebec, and the Prairie Provinces. The Northwest Territories have less potential and less direct sunlight because of their higher latitude. Ontario is scheduled to build one of the world's largest solar farms to be located in Sarnia and Sault Ste. Marie, that together will produce 60 MW – enough electricity for 9000 homes. Financially, however, electricity generation with photovoltaics is more costly than producing it with fossil fuels, hydro, or nuclear power. Accordingly, these solar farms must rely on government subsidies with a view to developing a solar-based industry.

Currently, solar energy technologies in Canada tend toward non-electrically active solar system applications for space heating, water heating, and drying crops and lumber. In 2001, there were more than 12 000 residential solar water heating systems and 300 commercial/industrial solar hot water systems in use. Though these comprise a small fraction of Canada's energy use, government studies indicate that they could make up as much as 5% of Canada's energy needs by 2025.

Geographically speaking, most of Canada's population is concentrated along a line that stretches about 100 miles (160 km) north of the American border, so that many regions north of this line tend to be sparsely populated and difficult to access. Hence, to power remote homes, telecommunications equipment, oil and pipeline monitoring stations and navigational devices Canada *has* to rely on photovoltaic (PV) cells used either as stand-alone units or as off-grid-distributed electricity generation. Canada's PV market has grown rapidly, and several companies have emerged making solar modules, controls, specialized water pumps, high efficiency refrigerators, and solar lighting systems. Since the 1970s, Canada's government and industry have encouraged development of solar technologies for remote communities. The use of hybrid systems provides power 24 hours a day using solar power when available and other energy sources when not.

Renewable Energy – Biomass Energy

Biomass energy is generated from burning plant material consisting mostly of agricultural and milling byproducts such as grain and wood pellets. Energy from biomass has been said by some folks to be carbon neutral in that the amount of carbon dioxide released when burning biomass equals the amount of CO_2 the plants absorbed during their growth (however, see Chapter 12). One of the best ways to use biomass is to mix it with coal before being fed into a furnace – this has several beneficial effects in that the amount of coal is thereby reduced, in addition to several environmental benefits that include (i) reduced CO_2 emissions, (ii) reduced sulfur emissions, and (iii) reduced mercury emissions. In fact biomass can, in principle, replace most of the coal being used in coal-fired electricity generating plants.

Biomass energy accounts for 540 PJ (petajoules) of energy use and provides more of Canada's energy supply than coal (for non-electrical generation applications) and nuclear power. It accounts for 5% of secondary energy use by the residential sector and 17% of energy use by the industrial sector, mostly by the forest industries. With lumber and pulp and paper, forestry accounts for 35% of Canada's total energy consumption with the forest industries meeting more than half of this demand themselves with self-generated biomass wastes. Canada's forest industries have increased their use of wood wastes that otherwise would have been burned, buried or else landfilled. The principal uses of wood wastes include firing boilers in pulp and paper mills for process heat, and the necessary energy for lumber drying.

Renewable Energy – Geothermal Energy

The largest conventional resources of geothermal power in Canada are located in British Columbia (BC), Alberta, and Yukon – these two provinces and territory also contain potential for enhanced geothermal systems (EGS). A 2007 estimate maintains that half of British Columbia's electricity needs could be met with geothermal energy. There are some 30 000 earth-heat installations that provide space-heating to Canadian residential and commercial buildings. The most advanced project under examination is a hybrid geothermal-electricity facility in British Columbia that could produce 100–300 MW power.

Renewable Energy – Sea Power

Until now, only tidal energy has seen some limited commercial development in Canada in a relatively short period of little over a decade – much research remains to be done on this potential energy source.

Detailed studies of the Bay of Fundy tidal-power resource have concluded that the most efficient development scheme would generate power for about 5 hours, twice daily, on the ebb tide. The most cost-effective project is a site in the Minas Basin (see Figure 37) at the mouth of Cobequid Bay in the upper reaches of the Bay of Fundy. The project would have a capacity in excess of 5300 MW, an amount equal to the entire 1980 installed generating capacity of the Maritime power systems. The cost of the project isn't cheap, however; it is estimated at about $7 billion.

Canada and Energy – Doing More

Canada's relationship to energy is unique by virtue of its size, its climate, its energy endowment, and its economy – its energy is singularly Canadian. According to the Canadian Center for Energy Information, energy ranks 4th as a contributor to Canada's GDP with nearly $82 billion in May 2011 (6.5% of total).

Figure 37 The Minas Basin, situated in Nova Scotia, Canada (photo by Chris Gotman/Valan Photos).

Compared to its global peers, Canada is an important energy provider depending, as it were, on the rare and useful balance between its relatively small population and the wealth of its energy resources. Canada's strength originates from the unique character of its energy production and consumption. Its geography and climate shape its consumption. Because of the size of its population relative to its geographic size, almost 30% of Canada's energy goes to fuel its transportation sector, and because of its cold northern climate, upwards of 40% of all energy produced is for heating.

Future energy use in Canada will continue to be shaped by the way it produces and consumes energy. With its expanding population and its growing economy, no doubt energy consumption will increase that will be offset by energy efficiency initiatives featured in the energy strategies adopted by Canada's provincial and territorial governments. In this regard, British Columbia (BC) continues to invest in wind power with its approval of the new Tumbler Ridge Wind Energy project in northeastern BC that will generate enough power to provide electricity for about 18 000 homes. As well, Nova Scotia's COMFIT program supports locally based renewable energy projects as part of its 2010 Renewable Energy Plan – in 2010 it approved five new projects.

Clearly, Canadians and their governments are doing more. Changes to the way Canada's citizens use energy will require an understanding of the country's energy mix, a complex array of sources and uses that will continue to determine Canada's global role.

12
Energy USA

Come forth into the light of things, let nature be your teacher.

William Wordsworth

Before he left office, President Nixon was quoted as saying that *gasoline will never exceed $1.00 per gallon*. Not to be outdone, President Carter stated that *the United States will never again import as much oil as it did in 1977*. In hindsight, they've both been proven wrong. Carter created the Department of Energy in 1977 as an off-shoot of the Atomic Energy Commission. Since those days much has happened in the United States on the energy front with several policies enacted that focused on energy conservation and energy development through government subsidies and incentives for renewable and non-renewable energy sources alike. These incentives are continued under the Obama Administration with a focus on the renewable sources.

Primary Energy Resources

At its birth in 1776, the United States–then consisting of the 13 colonies–used timber as the main energy source, mostly for heating purposes, which was subsequently replaced by coal in the nineteenth century because of ease of access to and ease of transportation of this fossil fuel. Soon thereafter (1916), coal had to compete with natural gas. However, coal provided the bulk of the United States energy demand well into the twentieth century, until it too was replaced by oil in the early 1900s, the transition due mostly to the development of the automobile. By 1950, oil consumption exceeded that of coal as a major source of energy, due in large part to the abundance of oil found in California, Texas, and Oklahoma, and in neighboring countries Canada and Mexico. However, coal continues to be used in many industrial processes, even as we speak.

Figure 38 illustrates the production of primary energies by sources for the period 1949–2010 in quadrillion BTUs (1 quadrillion = 1000 trillion; BTU = British thermal unit; 1 BTU = 1.055 kJ). The 1960–2000 period saw the production of coal increase while production of crude oil tended to decrease at the onset of the first oil crisis in the early 1970s–production of natural gas increased during this period,

Powering Planet Earth: Energy Solutions for the Future, First Edition. Nicola Armaroli, Vincenzo Balzani, and Nick Serpone.
© 2013 Wiley-VCH Verlag GmbH & Co. KGaA. Published 2013 by Wiley-VCH Verlag GmbH & Co. KGaA.

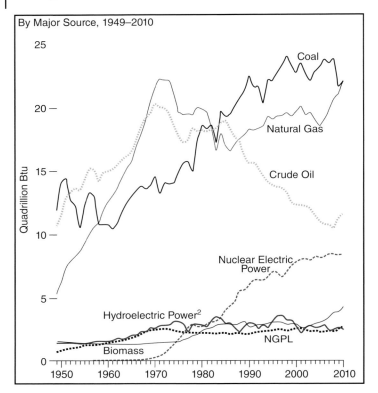

Figure 38 Production of primary energy by sources for the decades 1950 to 2010. Source: United States Energy Information Administration – Annual Energy Review 2010).

peaking in the 1970–1980 decade. Taken together, production of conventional fossil fuels remained essentially constant from 1970 to 2010 in the United States.

Nuclear electricity saw significant increases in the three decades spanning 1970–2000, reaching constancy in the first decade of the twenty-first century. By contrast, both hydroelectric power and production of NGPL (natural gas plant liquids) have remained fairly constant in the last four decades, whereas biomass has become an alternative energy source, showing some increase, particularly in the last decade.

The production of primary energies by sources for 2010 is portrayed in Figure 39. Geothermal, solar and wind power are the minor energy sources while natural gas and coal are the major sources of primary energy standing at 22 QBTUs (quadrillion BTU units) with crude oil not far behind at 12 QBTUs, followed by nuclear. Contrary to the 2010 picture, in 2007 the largest source of the United States' energy was oil (40%), followed by natural gas (24%) and coal (23%) with the remaining 15% coming from nuclear power, hydroelectric, and renewable sources.

The G. W. Bush Administration provided substantially more subsidies to fossil fuels ($72 billion) than to renewable energy sources ($29 billion). After all, Bush was a Texas oil man.

At times, the United States has not been averse to using its energy policy as a means to pursue some international goals, such as influencing the economy of

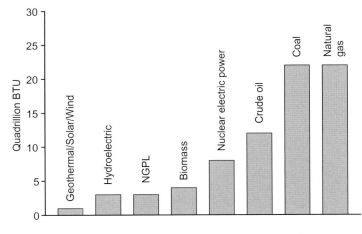

Figure 39 Production of primary energy by sources for 2010 in the USA. Data source: United States Energy Information Administration–Annual Energy Review 2010.

the former Soviet Union–basically bankrupting it by manipulating oil prices, which, in part, ultimately led to its dissolution.

Who can forget the 1973 oil crisis–at least those that lived through it? Who can forget the long lineups at the gas stations?

Many people felt the crisis was created by the major oil companies so that they could drive up the cost of crude oil, and by extrapolation their bottom line. Gone were the pre-1973 days when for a few pennies you could buy 1 United States gallon of gasoline. Unfortunately, the oil crisis also affected other economies–it was also deeply felt in Canada. Suddenly, the costs of buying real estate (a house) nearly doubled, nay, even tripled in some cases, and this in a matter of a few months (early 1974).

Several restrictions were imposed soon thereafter on the American consumer:

1) A national maximum speed limit of 55 mph (i.e., 88 km per hour) to reduce oil consumption.
2) Standards were enacted to downsize automobile categories.
3) Year-round daylight saving time was instituted.

In addition, the crisis also led to the search for alternative forms of energy and for diversifying oil supply sources. With regard to item 2, the 2007 Energy Independence and Security Act under the G. W. Bush Administration imposed an average gas mileage of 35 miles per gallon for cars by 2020. Obama's Administration encouraged the further development of plug-in electric cars for the near future and hydrogen-fueled cars by 2020.

Transportation, industry, and domestic sectors are the major users of about 84% of the energy that the United States gets from fossil fuels (Figure 40), with the remaining fraction coming primarily from hydro and nuclear installations. Although Americans represent only about 5% of the world's population, they consume 26% of the world's available energy and produce 26% of the world's

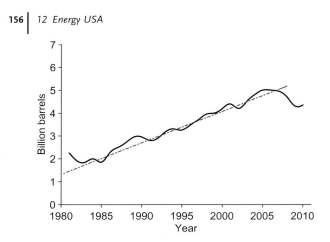

Figure 40 United States imports of crude oil and petroleum products (billion barrels) for the period 1980–2010. Note the increase in imports by nearly a factor of five over the period. Source: the United States Energy Information Administration.

industrial output. The United States accounts for 22% of oil consumption, yet it produces only 9% of the world's oil supply.

The known oil reserves of the United States account for only 1.9% of the world's reserves – that is, about 31 billion barrels (4.9 billion cubic meters). The United States is a voracious energy consumer. To satisfy this thirst for energy, nearly all of Canada's energy exports go to the United States, making Canada the largest source of United States oil imports. Canada is also the major source of United States imports of natural gas and electricity (the latter mostly from hydro sources).

America's imports grew from 10% in 1970 to 65% by the end of 2004. At the current rate of unhindered rise in imports, the United States will likely have to rely on about 70–75% of foreign oil by 2025. However, the Energy Information Administration (USEIA) projects that United States oil imports will remain flat and that consumption will grow, yet net imports are expected to decline to 54% of United States oil consumption by 2030.

On a per capita basis, estimates by the Central Intelligence Agency reported in the *CIA–The World Fact-book of 2008* show that America's daily oil consumption is twice that of the European Union, whose population is significantly greater. Not only are automobiles the single largest consumer of oil in the United States (40%), they are also the source of 20% of the nation's greenhouse gas emissions. Regrettably, the United States Congress failed to ratify the Kyoto Protocol, preferring instead to let the market drive CO_2 reductions to mitigate global warming.

Coal – Supply and Demand

Domestic sources of coal have been the primary source of energy in the United States from 1885 to 1951, with crude oil and natural gas vying for that role in the

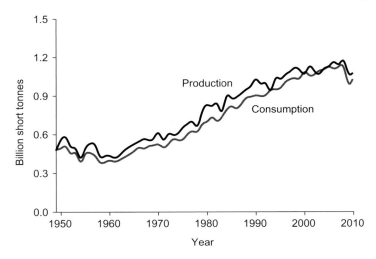

Figure 41 Production and consumption of coal in billion short tons in the United States since 1949. Data source: Annual Energy Review of October 19, 2011 – United States Energy Information Administration.

period 1952–1982. Coal took top spot in domestic resources in 1982 and then again in 1984, retaining top position ever since. The United States is self-sufficient in this energy resource, which is expected to last for at least a few hundred years more. The trend in the production and consumption of coal in the United States has been on the increase ever since the 1950s – nearly tripling by 2006 – in part due to the doubling of the United States population during this period (Figure 41). Per capita consumption has been on the decrease since 1978.

Coal power accounted for about 45% among other sources in the production of electricity in 2009 (see Figure 42), with the utilities accounting for more than 90% of purchases of United States domestic coal. For this purpose, they burn nearly 1 billion tons of coal annually to feed coal-fired electric power stations. Historically, more than 90% of coal consumed in 2006 was used to generate electricity compared with about 19% back in 1950.

In 2009, the United States had 1436 coal-powered units at electrical utilities, whose total nominal capacity was about 339 GW compared to 1024 units at a nominal capacity of 278 GW in 2000 – in that year the average production of electricity from coal was 224.3 GWh. The quantity of coal consumed in 2006 was 1.03 billion short tons (or 0.931 billion tonnes), which represented 92.3% of coal consumed for electricity generation.

Combustion of such large quantities of coal has caused great concern with regard to its effect on climate change since coal is the largest source of CO_2 emissions. Emissions from electricity generation accounted for the largest share of United States greenhouse gases: 38.9% in 2006, with the transportation sector accounting for 31% of CO_2 emissions. Although coal power only accounted for 49% of the United States electricity production in 2006, it was responsible for

2009 U.S. Electricity Generation by Source

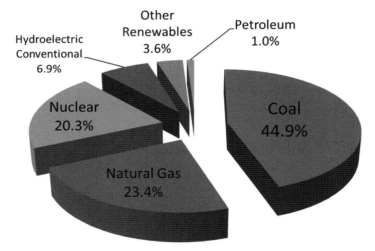

Figure 42 Sources of electricity in the United States in 2009 (by Daniel Cardenas – see http://www.eia.doe.gov/cneaf/electricity/epm/table1_1.html).

83% of CO_2 emissions from electricity generation that year – however, this did not include additional greenhouse gas (GHG) emissions from coal, including methane from coal mines, emissions from coal transport, other GHG emissions (e.g., particulates or black carbon), and carbon and nitrous oxide (N_2O) emissions from land transformation in the case of mountain top removal coal mining. This led environmentalists to call for a moratorium on all coal consumption unless carbon emissions could be captured and sequestered. Some people suggested that the best technology to sequester carbon was to leave the coal in the ground.

At present, the cleanest operational coal-fired electricity generation technology in the United States is the Integrated Gasification Combined Cycle (IGCC) method, with another methodology (FutureGen) being examined for the possible sequestration of IGCC-CO_2 emissions underground.

In addition to CO_2 emissions, combustion of coal is also responsible for the release of other no less significant pollutants into the atmosphere with no less harmful effects on the environment. For example, emitted byproducts from coal-fueled plants have been linked to acid rain – 86 coal powered plants with a capacity of about 107 GW (9.9% of total United States electric capacity) emitted nearly 5.4 million tons of sulfur dioxide (SO_2) in 2006 or about 29% of SO_2 emissions from all United States sources.

Another harmful pollutant released into the environment is mercury (Hg). According to the United States Department of Energy (USDOE), United States coal-fired electricity-generating power plants emitted an estimated 48 tonnes of Hg in 1999, the largest source of man-made mercury pollution in the United States. In the period 1995–1996, this accounted for almost 33% of all mercury emitted into the air by human activity in the United States alone – in addition, about 13% of the Hg was emitted by coal-fired industrial and mixed-use commercial boilers, and 0.3% by coal-fired residential boilers, bringing the total United States mercury pollution from coal combustion to 46% of United States man-made mercury sources. By contrast, China's coal-fired power plants emitted an estimated 200 tons of Hg in 1999, or about 38% of Chinese human-generated mercury emissions (45% being emitted from non-ferrous metals smelting). The use of activated carbon (a form of carbon processed to be riddled with small, low-volume pores that increase the surface area available for adsorption) can attenuate mercury emissions from power plants.

Natural Gas – Supply and Demand

Production and consumption of natural gas increased by a factor of four in the decades between 1959 and 1970 to about 20 trillion cubic feet (about 566 billion cubic meters), then declined gradually to stabilize in 1986 (Figure 43). Since then, the United States has imported an increasing quantity of its need of natural gas – 90% of its imports were supplied by Canada (about 3 billion cubic feet or 85 billion cubic meters).

Consumption of natural gas in 2008 was about 23.2 trillion cubic feet (660 billion cubic meters), while domestic production was only 20.6 trillion cubic feet (580 billion m^3). Other foreign supplies of natural gas were delivered by tankers as LNG (liquefied natural gas) from five other countries.

Figure 44 shows the domestic sources of natural gas in the United States. In 2007, Texas (30%), Wyoming (10%), Oklahoma (9%) and New Mexico (8%) were the largest gas-producing states – another 14% came from offshore in the Gulf of Mexico. Recent developments in hydraulic fracturing technology, together with horizontal drilling, have increased America's interest in shale gas, which has led to greater natural gas reserves, about 35% higher in 2008 than in 2006 due largely to discoveries of shale gas sources. Figure 45 illustrates the 2010 estimates by the

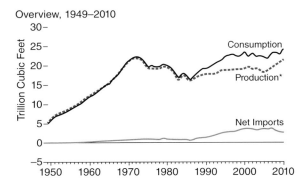

Figure 43 Production and consumption together with net imports of (dry) natural gas during the period 1949–2010. Data source: the United States Energy Information Administration – Annual Energy Review 2010.

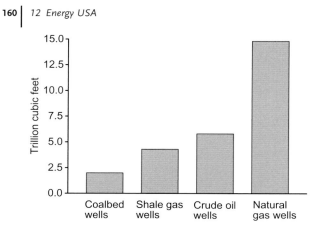

Figure 44 Domestic sources of natural gas as gross withdrawals by well type in the United States for the year 2010. Data source: United States Energy Information Administration–Annual Energy Review 2010.

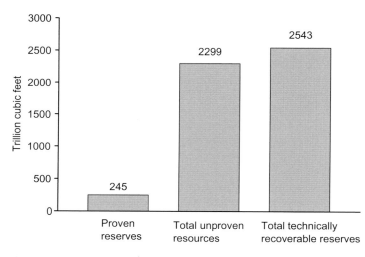

Figure 45 2010 estimates of natural gas reserves in the United States. Data source: United States Energy Information Administration–Annual Energy Review 2010.

United States Energy Information Administration on the reserves of natural gas resources, although it would appear that not everyone in the EIA agreed with the optimistic projections of reserves.

Shale gas resources have become a rapidly increasing source of natural gas in the United States led by novel technologies to extract natural gas, which have been mostly responsible for stopping the decline of supplies from conventional sources. Economic success of shale gas in the United States since 2000 incited Canada to explore shale gas resources (see Chapter 11)–more recently, interest in shale gas exploitation has also been envisaged by Australia and some countries in the European Union and Asia.

Shale gas wells produced 300 billion cubic feet (8.5 billion cubic meters) in 1996, or about 1.6% of United States gas production. By 2006, production nearly quadrupled to an annual 1100 billion cubic feet (31 billion cubic meters) or 5.9% of United States gas production. There were 14 990 shale gas wells in the United States in 2005 – in 2007 there were an additional 4185 wells.

By 2008, United States shale gas production came to 2.02 trillion cubic feet (57 billion cubic meters), a 71% increase from the previous year, increasing by another 54% to 3.11 trillion cubic feet (88 billion cubic meters) in 2009. Proven United States shale gas reserves increased by 76% to 60.6 trillion cubic feet (1.72 trillion cubic meters) at the end of 2009. In its Annual Energy Outlook for 2011, the United States Energy Information Administration (EIA) more than doubled its estimate of technically recoverable shale gas reserves in the United States, to 827 trillion cubic feet (23.4 trillion cubic meters) from 353 trillion cubic feet (10.0 trillion cubic meters). Shale gas production is expected to nearly triple from 14% of total United States gas production in 2009 to 45% by 2035.

> *The development of shale gas is expected to significantly increase United States energy security and help reduce greenhouse gas pollution.*
> White House, Office of the Press Secretary, 17 November 2009

It has been proposed by some people that coal-powered electricity installations be replaced by natural gas-fired power plants as the latter emit less greenhouse gases – they could also serve as a backup power source for wind energy.

Nuclear Power

Historical Notes

The use of nuclear power has faced strong opposition, not only from environmentalists but also from various organizations in the United States such as the United Auto Workers Union. One of the first nuclear reactors to face opposition was the Fermi-1 facility built in 1957 near Detroit, Michigan. The first commercially viable nuclear power plant in the United States was planned to be built in 1958 at a site located just north of San Francisco, but strong opposition to the project by local citizens for nearly 6 years ultimately led to its abandonment. Attempts to build a nuclear power plant in Malibu, California, suffered a similar fate.

Nuclear accidents in the 1960s involved a small test reactor in Idaho Falls in January 1961 and the partial meltdown of a nuclear electricity generating station in Michigan in 1966. These two incidents provided ammunition to antinuclear activists. Though environmentalists see some advantages of nuclear power in reducing air pollution, they're quite critical of nuclear technology on other grounds such as nuclear accidents, nuclear proliferation, the high cost of nuclear power plants and radioactive waste disposal, and now nuclear terrorism.

The nuclear industry has failed to address the nuclear waste issue. Spurred by the 1979 Three-Mile Island nuclear accident it became evident that nuclear power would not grow as once believed. Eventually, more than 120 nuclear projects were withdrawn and the construction of new reactors was abandoned.

Vice-President Al Gore had this to say on the historical record and reliability of nuclear power in the United States:

> *Of the 253 nuclear power reactors originally ordered in the United States from 1953 to 2008, 48% were canceled, 11% were prematurely shut down, 14% experienced at least a one-year-or-more outage, and 27% are operating without having a year-plus outage. Thus, only about one fourth of those ordered, or about half of those completed, are still operating and have proved relatively reliable.*

Harvard University-educated environmental scientist and writer Amory B. Lovins also commented on the history of the nuclear option for the United States:

> *Of all 132 United States nuclear plants built (52% of the 253 originally ordered), 21% were permanently and prematurely closed due to reliability or cost problems, while another 27% have completely failed for a year or more at least once. The surviving United States nuclear plants produce ~90% of their full-time full-load potential, but even they are not fully dependable. Even reliably operating nuclear plants must shut down, on average, for 39 days every 17 months for refueling and maintenance, and unexpected failures do occur too.*

A February 11, 1985, article in *Forbes Magazine* questioned the overall management of the nuclear power program in the United States:

> *The failure of the U.S. nuclear power program ranks as the largest managerial disaster in business history, a disaster on a monumental scale . . . only the blind, or the biased, can now think that the money has been well spent. It is a defeat for the U.S. consumer and for the competitiveness of U.S. industry, for the utilities that undertook the program and for the private enterprise system that made it possible.*

According to a report by the Nuclear Regulatory Commission, the Three Mile Island accident was the most serious in United States commercial nuclear power plant operating history, even though it caused neither casualties nor injuries to personnel or to the local citizenry. A subsequent lengthy (13 years) investigation of this accident involving 32 000 people found no adverse health effects which might have been linked to the accident. The United States is the world's largest supplier of commercial nuclear power.

Present Situation

As of 2011, nuclear power in the United States was provided by 104 commercial reactors, 69 of which are pressurized water reactors and 35 are boiling water reac-

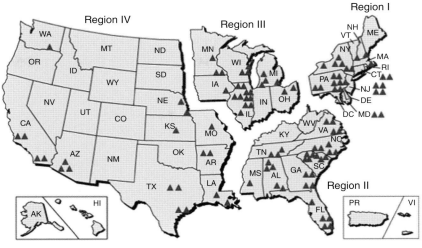

Figure 46 Nuclear Regulatory Commission regions and locations of nuclear reactors; source: http://www.nrc.gov/info-finder/reactor.

tors, licensed to operate at 65 nuclear power plants that produce a total of approximately 790 TWh of electricity, that is, 19.2% of the United States total electric energy generation in 2011.

Currently, demand for nuclear power has lessened to such an extent that some companies have even withdrawn their building lease applications. Construction of two nuclear plants with a total of four reactors was projected in the mid-1970s, but the only nuclear power plant under construction is in Tennessee, which begun in 1973 and will likely be completed in 2012. However, in 2012, the Obama Administration gave the go-ahead to build two nuclear power plants.

Following the Fukushima disaster of 2011, the United States Nuclear Regulatory Commission launched a comprehensive safety review of all the nuclear reactors across the United States (see Figure 46) at the request of President Obama, who nonetheless supported the expansion of nuclear power in the United States, despite the crisis in Japan. Public support for nuclear power in the United States has dropped to 43%, somewhat less than soon after the Three Mile Island accident, in spite of the many technical studies that have asserted a low probability of a severe nuclear accident. Numerous surveys have shown that the public remains *very deeply distrustful and uneasy about nuclear power*.

Some commentators have even suggested that the public's consistently negative ratings of nuclear power reflect the nuclear industry's unique connection with nuclear weapons.

Despite some opposition, however, a series of Gallup polls, taken between 1994 and 2009, found that support of Americans for nuclear energy was mixed, hovering around 46% to 59%, with significantly different opinions between genders, income groups, and political affiliation (Democrats vs Republicans). Any renewed interest in nuclear power in the last decade has been facilitated in part by the United States Government's Nuclear Power 2010 Program, which coordinates efforts for building new nuclear power plants.

Nuclear Renaissance

The prospect of a *nuclear renaissance* has revived the debate regarding the nuclear waste issue. Although there appears to be an *international consensus on the advisability of storing nuclear waste in deep underground repositories*, no country has yet found such a site. Nuclear proliferation concerns induced the Obama Administration to disallow reprocessing of nuclear waste within America's borders.

> *The nuclear renaissance is looking small and slow at the moment.*
> Matthew Wald, *New York Times*, September 2010

As of March 2009, the Nuclear Regulatory Commission was considering 26 applications for new nuclear power reactors with an additional 7 expected – 6 of these were ordered. However, it's unlikely that these reactors will be built in view of recent events in Japan. The operator of 17 nuclear reactors in the United States was scheduled to cancel or delay construction of 2 new reactors unless it obtained government loan guarantees because of the high costs and risks in building new reactors. On February 2010, President Obama announced loan guarantees for two new reactors which would be the first nuclear power plants with the go-ahead to be built in the United States since the mid-1970s.

> *The reactors are just the first of what we hope will be many new nuclear projects.*
> Carol Browner, Director,
> White House Office of Energy & Climate Change Policy.

Contrary to the 1994–2004 Gallup poll, a recent Washington Post-ABC poll conducted in April 2011 found that 64% of Americans oppose the construction of new nuclear reactors. At the same time, 45 groups and individuals have petitioned the Nuclear Regulatory Commission to suspend all licensing and activities at 21 proposed nuclear reactors until such time that the Commission completes its study of the Fukushima incident. In the aftermath of this catastrophic event, several issues will impact the operators of nuclear power plants in the United States, namely:

1) Increased safety costs of currently operating plants and new nuclear power installations.

2) License extensions of current reactors will be subject to additional scrutiny. This will likely involve 60 of the 104 currently operating United States reactors.

3) On-site storage, consolidated long-term storage, and geological disposal of spent fuel are likely to be re-evaluated because of the Fukushima storage pool experience.

In 2011, the London-based HSBC Bank stated that with Three Mile Island and Fukushima as a backdrop, the United States public may find it difficult to support major nuclear new build, and the Bank also expected that no new plant extensions would be granted. It further expected that the clean energy standard under discussion in the United States Congress would emphasize gas, renewable energies, and process efficiency. And on this, Mark Z. Jacobson of Stanford University remarked that:

> *If our nation wants to reduce global warming, air pollution and energy instability, we should invest only in the best energy options. Nuclear energy isn't one of them.*

Water Usage in Nuclear Reactors

Even though studies have shown no significant environmental impact of once-through cooling systems, their use has recently been seriously questioned because of possible damage to the environment from increased warm/hot water discharged to local surface waters, together with possible contamination from leaked radioactive substances. Regulations of the United States Environmental Protection Agency (USEPA) favor water recirculation systems, and require that older nuclear power plants replace existing once-through cooling systems with new recirculation systems. Germane to this, an Associated Press study of 2008 on the 104 nuclear reactors in the United States reported that

> . . . *24 (reactors) are located in areas experiencing severe drought levels, and all but two are built near lakes and rivers and so they rely on underwater intake pipes to draw billions of gallons of water to cool the reactors and as a source of steam for the plants' turbines.*

During the 2008 southeast drought, reactor output was either reduced to lower operating power levels or else the reactors were forced to be shut down altogether.

Plant Decommissioning

Decommissioning a nuclear reactor is a costly enterprise in terms of both energy required – as much as 50% more than the energy needed for the original construction – and financial resources needed (from $ 300 million to $ 5.6 billion) for long periods of time (about 50 to 100 years) after the facility has finished generating its last useful electricity – both the nuclear reactors and the uranium enrichment facilities must also be decommissioned to return the facility and its infrastructure to safe levels for other uses. Decommissioning reactors is not only very expensive – especially those that have experienced a serious accident – but can also be time-consuming, dangerous to workers, hazardous to the natural environment, and could present opportunities for human error, accidents, and even sabotage. Thirteen nuclear reactors have either been shut down permanently or are in the process of being decommissioned.

Renewable Energy

Every nation on spaceship Earth wishes for an energy mix that is secure, reliable, improves public health, protects the environment, addresses climate change, creates jobs, and provides technological research. The one energy resource that meets these requirements is the renewables, which as we have noted in several places in this book comprise a broad, diverse array of technologies that include solar photovoltaics, solar thermal power plants and heating/cooling systems, wind farms, hydroelectricity, geothermal power plants, ocean power systems, and not least the use of biomass.

The United States has one of the best mixes of renewable energy resources in the world – they have the potential to meet a rising and significant share of the nation's energy demand. In the first six months of 2011, renewables accounted for 14.3% of the domestically produced electricity, while hydroelectricity was the largest producer of renewable power. In 2009, the United States was the world's largest producer of electricity from geothermal, solar, and wind power, trailing only China in the total production of renewable energy.

Total renewable energy consumed in 2009 amounted to about 8%, while approximately 10% of the United States electricity was produced from renewable sources. Hydroelectricity accounted for 67% (or about 248 GW) of the country's renewable energy, the rest coming from other renewable sources (see Figure 47 for 2010).

Increases in wind, solar, and geothermal power were expected to nearly double production of renewable energy by 2012, most coming from wind power (Figure 48). Most cars in the United States now run on gasoline blends containing upwards of 10% ethanol fuel produced from treated corn, and this figure is expected to increase as car manufacturers are now making cars that can use gasoline blends higher in ethanol.

Obama's Administration planned to invest some $150 billion within 2025 to catalyze private efforts in building a clean energy future. Specifically, the plan

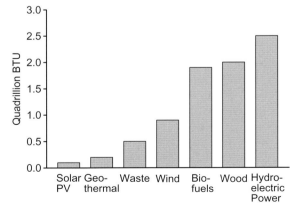

Figure 47 Renewable energy consumption in quadrillion BTUs from various sources for 2010. Data from the United States Energy Information Administration.

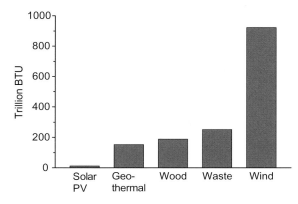

Figure 48 Power generated from various sources for 2010. Data from the United States Energy Information Administration.

called for renewable energy to supply 10% of the nation's electricity by 2012, rising to 25% by 2025.

> *We know the country that harnesses the power of clean, renewable energy will lead the 21st. century. . . . Thanks to our recovery plan we will double this nation's supply of renewable energy in the next three years . . . It is time for America to lead again.*
> President Barack Obama, Joint address to Congress, 2009

In September 2011, investment in renewable energy suffered a setback when Solyndra declared bankruptcy owing to plummeting silicon prices that made it unable to compete with more conventional solar photovoltaic panels. This once-promising solar energy manufacturer of cylindrical panels of CIGS (copper indium gallium diselenide) thin-film solar cells located in California had earlier received $535 million in federal loans from the somewhat embarrassed Obama Administration.

Renewable Energy—Wind Power

Over the last few years, wind power has experienced a near-exponential growth in the United States, although 2010 experienced fewer new constructions compared with the previous year (Figure 49). In 2008, installed capacity increased by 50% compared with 2007, while compared the world's average growth rate that year was about 29%. The United States is second only to China in installed capacity of wind power, which is also experiencing rapid growth worldwide. A 2008 report by the United States Department of Energy foresaw wind power as supplying 20% of all United States electricity by 2030, including a contribution of 4% to the nation's total electricity from offshore wind power. Reaching this goal, however, will require significant advances in cost, performance and reliability.

A study by the National Renewable Energy Laboratory (NREL), published in February 2010, indicated that the contiguous United States (excludes Hawaii and

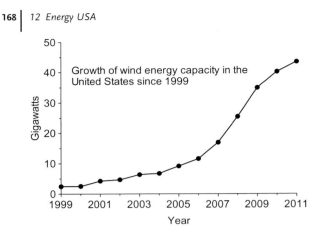

Figure 49 Wind energy capacity growth in gigawatts in the United States since 1999. Data from the United States Energy Information Administration.

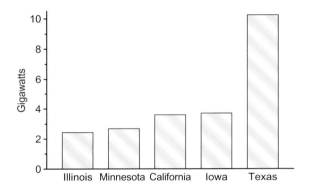

Figure 50 Graphic illustrating the five states with the greatest installed wind energy capacity in gigawatts in the United States (2011). Data source: the United States Energy Information Administration.

Alaska) had the potential to install about 10.5 TW of onshore wind power. This capacity could generate 37 petawatt-hours (PWh) annually, an amount nine times greater than 2010 total electricity consumption. Alaska and Hawaii also possess large wind resources. In fact, a quarter of the United States landmass has winds strong enough to generate electricity at the same price as natural gas and coal.

As of September 2011, the cumulative installed capacity of wind power in the United States stood at 43.5 GW; 2.3% or about 94.7 GWh of generated electricity in 2010 came from wind power. With 9.73 GW of capacity, Texas is ahead of any state, followed by Iowa with 3.67 GW, thus establishing Texas as the leader in wind power development, followed by Iowa and California (Figure 50).

There are at present some 90 projects under construction for an additional capacity of 8.48 GW that are generating tens of thousands of jobs and billions of dollars of economic activity – in particular, it is revitalizing the economy of rural communities by providing a steady income stream to farmers with wind turbines on their land. Farmers are typically paid $3000–5000 annually in royalties from the local utility for providing a site for a single, large, advanced-design wind turbine that occupies a quarter of an acre (1000 square meters) of their land, which would normally produce 40 bushels of corn worth $120 or, in ranch country, beef worth perhaps $15. Clearly, this is a win-win situation, especially for the farmer.

In terms of cost, consider this: in the early 1980s, the cost of wind-generated electricity in California was $0.38 per kilowatt-hour—since then it has dropped to $0.04 (or less) at the best wind sites; some long-term supply contracts on wind power are looking at costs of about $0.03 per kilowatt-hour making wind energy one of the world's most economical sources of electricity.

Not everyone in the United States is pleased with wind farms, however, which many people still consider an eyesore on the landscape. Such concerns must be weighed against the threats posed by other energy sources to climate change and to the needs of the community. Nonetheless, if an appropriate site selection is chosen and environmental problems are addressed, worldwide experience has shown that, when a community is consulted and directly involved in wind farm projects, these factors help enormously in increasing a community's approval of such installations.

In addition to large onshore wind resources, the United States has very large offshore wind energy resources that it could exploit because of consistent strong winds along the United States coastlines. Here too, some coastal residents oppose offshore wind farms because of fears of possible damage to marine life, the environment, electricity rates, aesthetics, and most of all recreational activities such as fishing and boating. However, studies have shown that through careful planning when locating wind farms and positioning the wind turbines at some distance from shore, except for some initial disturbance of the construction phase, marine life and recreation facilities are not expected to be affected. Despite their fears, local residents do recognize the potential benefits of wind farms in improved electricity rates, air quality, and job creation.

A report released by NREL in September 2010 showed that the United States has 4.15 TW of potential offshore wind power capacity, an amount 4 times greater than the country's 2008 installed capacity from all sources (1.01 TW). However, as of 2011, the United States had no offshore wind farms, although there are exploratory projects for wind power production on the Outer Continental Shelf offshore from New Jersey and Delaware. A national strategy directed at offshore wind power expects to produce 10 GW in 2020 and 54 GW in 2030. A robust United States offshore wind industry could generate tens of thousands of additional jobs and billions of dollars of economic activity.

Another NREL report on the current state of the United States offshore wind industry concluded that development of the nation's offshore wind resources can provide many potential benefits, and, with effective research, policies, and commitment, offshore wind energy can play a vital role in future United States energy markets.

Renewable Energy – Solar Thermal Power

The southwestern region of the United States is one of the world's best areas for solar radiation (insolation). Compared to other regions, the Mojave Desert receives nearly twice as much sunlight, so that installing solar power plants in the desert makes good sense and provides cleaner alternatives to traditional fossil fuel power plants – no emissions, and no fuel consumed other than sunlight. Contrary to conventional coal and nuclear power plants that require long lead times to build, solar thermal power plants can be built in a few years with modular, readily available materials. Financing of such installations from private sources doesn't come easy, and so they often have to rely on government subsidies, or at the very least on loan guarantees, as solar electricity is not cost-competitive with other bulk power sources although it does mitigate the risk of fuel-price volatility.

According to the United States Department of Energy, more than 1.5 million homes and businesses used solar water heaters in 2006, representing a capacity of over 1000 MW of thermal energy generation. If 40% of existing homes had adequate access to sunlight, 29 million solar water heaters could be installed in the United States. Solar water heaters can operate in any climate, and reduce the need for conventional water heating by about two-thirds – payback time is about 4 to 8 years with electricity or natural gas savings, as experienced by Florida home owners who, according to the Florida Solar Energy Center, save on average 50–85% on their water heating bills compared to those who use electric water heaters.

Solar thermal power installations often take up large landmasses – up to about 5 to 10 square miles – and so it's not surprising to find these solar plants at some distance away from populated areas. And although they are large with respect to the electricity output, they use less land than hydroelectric dams (including the size of the lake behind the dam) and coal plants (including the amount of land required for mining and the excavation of coal).

The Solar Energy Generating Systems (SEGS) technology was pioneered in the United States with the Solar One installation in the Mojave Desert together with several others, making up the SEGS group of 9 solar thermal power stations whose total generating capacity is 354 MW, and making the system the largest solar thermal plant of any kind in the world.

Figure 51 illustrates one of the SEGS systems and its mode of operation. Briefly, the group of parabolic-shaped troughs (or collectors) amplifies the sun's radiation 30 to 60 times (that is 30 to 60 suns) its normal intensity (one sun) and focuses it onto receiver pipes positioned at the focal line of the troughs. This heats up the

Figure 51 Example of a solar electric generating system (SEGS) and its mode of operation: Solar collectors capture and concentrate sunlight to heat therminol – a synthetic oil – which then heats water to produce steam that is piped to an onsite conventional steam turbine generator to produce electricity, which is then transmitted over power lines. On cloudy days, the plant has a supplementary natural gas system to produce steam and then generate electricity. Source: http://www.nexteraenergyresources.com/pdf_redesign/segs.pdf.

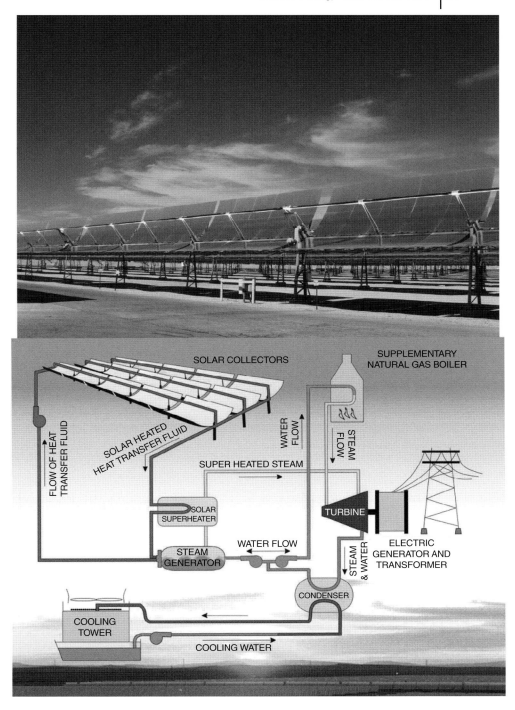

SOLAR COLLECTORS

SUPPLEMENTARY
NATURAL GAS BOILER

FLOW OF HEAT
TRANSFER FLUID

SOLAR HEATED
HEAT TRANSFER FLUID

WATER
FLOW

STEAM
FLOW

SUPER HEATED STEAM

SOLAR
SUPERHEATER

TURBINE

ELECTRIC
GENERATOR AND
TRANSFORMER

STEAM
GENERATOR

WATER FLOW

STEAM
& WATER

CONDENSER

COOLING
TOWER

COOLING WATER

synthetic oil that circulates through the pipes to 390 °C (735 °F), which is then pumped to a generating station and routed through a heat exchanger to produce steam that drives a conventional steam turbine to finally generate electricity. On cloudy days or after dark, the SEGS plants operate on natural gas that provides 25% of total output.

Another type of solar thermal power plants uses thousands of individual sun-tracking mirrors (heliostats) to reflect solar radiation onto a central receiver located on top of a tall tower. A consortium of United States utilities and the United States Department of Energy built the first two demonstrations of large-scale solar power towers in the California desert: Solar One and Solar Two, which have now been decommissioned. Solar One was operated between 1982 and 1988, and used water/steam as the heat-transfer fluid in the receiver – this presented several problems in terms of storage and continuous turbine operation. Accordingly, Solar One was upgraded to Solar Two (Figure 52), which operated from 1996 to 1999 and used molten salt to capture and store the sun's heat to a turbine/generator to produce electricity (ca. 10 MW power). The system operated smoothly through intermittent clouds and continued generating electricity long into the night.

The Nevada Solar One (not to be confused with the Solar One demonstration project in California) installation near Boulder City, Nevada, is spread over an area

Figure 52 Aerial view of the Solar Two facility, showing the power tower (left) surrounded by the sun-tracking mirrors. Source, http://earthobservatory.nasa.gov/Features/ RenewableEnergy/Images/solar_two.jpg.

of 400 acres and uses the technology that collects and stores the extra heat by putting it into phase-changing molten salts. Its nominal capacity is 64–75 MW, and its annual amount of avoided CO_2 emission is equivalent to 20 000 cars taken off the road. This project began operation in June 2007 following an investment of $266 million, with annual electricity production estimated at 134 GWh. Other solar thermal power stations have been built or are otherwise under construction, two in California (capacity 392 MW and 968 MW), one in Nevada (110 MW), and one in Arizona (280 MW).

The 392-MW Ivanpah Solar Power Facility, under construction 64 km from Las Vegas, Nevada, is the world's largest solar-thermal power plant project, occupying 5.6 square miles at a cost of about $1.6 billion. Once completed, the facility will deploy 347 000 heliostat mirrors that focus the sun's radiation onto boilers located on centralized solar power towers. The project is not without some controversy, however, with regard to its environmental impact.

Renewable Energy – Solar Photovoltaic

The Mojave Desert is definitely an excellent location to install photovoltaic generation of electricity because of the available insolation. A 230-MW photovoltaic project is under construction in the Antelope Valley area of the Western Mojave Desert, to be completed in 2013. It features an innovative utility-scale deployment of inverters with voltage regulation and monitoring technologies, which will enable the project to provide stable and continuous power.

The Nellis Solar Power Plant, located on 140 acres (57 hectares) of land leased from the United States Air Force (USAF) at the western edge of the base in Clark County, Nevada, was touted as one of the largest (size-wise) solar photovoltaic systems in North America – it was completed in December 2007 (Figure 53). This ground-mounted photovoltaic field employs an advanced sun tracking system wherein each set of solar panels rotates around a central bar to track the sun. The 14-MW system generates more than 30 GWh annually or about 82 MWh daily, with approximately 25% of the total power being used by the Air Force base.

An even larger photovoltaic power plant in North America is the 48-MW Copper Mountain Solar Facility in Boulder City about 40 miles from Las Vegas. Construction began in January 2010 and was completed on December 1st – a very short time indeed – at which time the facility began to generate electricity. It took more than 350 construction workers to install the 775 000 First Solar panels to power the plant on the 380-acre site. This moved the United States into the *top five* category when it came to large PV power plants – at the time, only Canada, Italy, Germany and Spain had bigger plants.

Yet another solar photovoltaic project – the Blythe Solar Power Project – that would have generated 968 MW power was proposed for construction at Riverside County, California. Originally the facility was to use a solar thermal parabolic trough design, but later it was switched to solar photovoltaic panels using the same

Figure 53 Photograph of the Nellis Solar Power Plant at Nellis Air Force Base; the panels track the sun in one axis. Source: http://www.nellis.af.mil/photos/media_search.asp?q=solar&btnG.x=0&btnG.y=0.

technology as used in rooftop installations – lower costs were the principal reason for the switch. Unfortunately, lack of financing to begin construction of the project forced the company to sell the project to a German enterprise, which is looking into forming a joint venture with an American firm to build the Blythe project. Five other no less significant solar projects are also being considered by the United States Department of the Interior to use federal lands. There are a total of 23 GW of utility-scale solar projects in the development pipeline in the United States, which if realized would make it the leading country in this field.

Renewable Energy – Geothermal Energy

As noted earlier in this book, geothermal energy is contained in underground reservoirs of steam, hot water, and hot dry rocks. Archeological finds have shown that geothermal resources were in use in the contiguous United States more than 10 000 years ago. The Paleo-Indians used geothermal hot springs for warmth, cleansing, and minerals.

As used at electric generating facilities, hot water or steam extracted from geothermal basins in the Earth's crust is supplied to steam turbines to produce electricity. Moderate-to-low temperature geothermal resources are usually used for direct-use applications such as district and space heating, whereas lower-temperature, shallow-ground geothermal resources are used by geothermal heat pumps to heat and cool buildings. Unlike wind and solar resources, which depend on weather, geothermal resources are available 24/7 (that is, 24 hours a day, 7 days a week).

Figure 54 Geothermal power plant at The Geysers near the city of Santa Rosa in northern California. The Geysers area is the largest geothermal development in the world. (Photograph by Julie Donnelly-Nolan, USGS.) Source: http://pubs.usgs.gov/gip/dynamic/geothermal.html.

Geothermal energy is an area of considerable activity in the United States, which in 2010 led the world in online capacity and generation of electricity from this energy source with 3.09 GW of installed capacity from 77 power plants. The largest group of plants in the world is located at The Geysers (Figure 54), located about 116 km (72 miles) north of San Francisco. In fact it is here that the first commercial geothermal power plant producing 11 MW of net power and delivering it to the United States utility grid began operation in 1960. This system still operates today with a total output of 750 MW. The Geysers system is now recharged by injecting treated sewage effluent that used to be dumped into rivers and streams—it is now piped to the geothermal field where it is converted into steam for power generation.

Geothermal power plants are concentrated mainly in the western states, and are the fourth largest source of renewable electricity after hydroelectric plants, biomass, and wind farms. An assessment of geothermal resources showed that 9 western states have a potential of providing over 20% of national electricity needs.

As of 2011, there were 43 geothermal electricity-producing plants in California—combined capacity, about 1800 MW. In addition, regions of Nevada, Oregon, Idaho, Arizona, and Utah are witnessing rapid geothermal development.

Though several small geothermal power plants were built in the late 1980s, as a result of high power prices, rising energy costs have stimulated new development.

The United States generates an annual average of 15 TWh of geothermal energy, comparable to burning annually some 25 million barrels of oil (i.e., 4 million cubic meters) or 6 million short tons of coal. At the current rate of development, geothermal production of electricity is expected to exceed 15 GW by 2025. With its current installed geothermal capacity, the United States is the world leader, with 30% of online total capacity. The future outlook for expanded production from conventional and enhanced geothermal systems is bright, as new technologies promise increased growth in locations previously not considered.

In this regard, the United States geothermal market is adding new projects to its development pipeline each year, and 2012 is not expected to be different. The industry entered the New Year 2012 with up to billions of dollars in planned investments – currently up to 5.7 GW are in the development stage. The Energy Policy Act of 2005 catalyzed much of the industry's activity in geothermal energy resources, as the Act made new geothermal plants eligible for full United States Government tax credit, previously available only to wind power projects and certain kinds of biomass – the Act further authorized and directed increased funding for research.

Renewable Energy – Biomass

We've talked much about biomass earlier in this book but never really described what biomass is made of, except in a very rudimentary way. Biomass is best described as material derived from living, or recently living organisms – for example, forest residues (dead trees, branches and tree stumps), yard clippings, wood chips, municipal solid waste, and plant matter specifically grown to generate electricity or produce heat.

Biomass also includes plant or animal matter that is converted into fibers or other industrial chemicals, including biofuels, whereas industrial biomass is grown from numerous types of plants, including miscanthus, switchgrass, hemp, corn, poplar, willow, sorghum, sugarcane, and a variety of tree species ranging from eucalyptus to palm oil. While debate regarding the net carbon neutrality continues, a key difference is the relatively short carbon recycle period of grown biomass (several years or decades) versus the millions of years that it took to convert carbon into fossil fuels.

As an energy source, biomass can be used either directly or converted into other forms of energy such as methane gas or biofuel (ethanol) for transportation. Biodiesel can also be produced from leftover food products such as vegetable oils and animal fats. Rotting garbage and agricultural and human waste release methane gas, also called *landfill gas* or *biogas*. Used directly, biomass can generate electricity through combustion and can heat and/or cool homes and buildings.

Biomass for electricity production typically depends on the regions where it is used – for instance, forest by-products such as wood residues are popular in the United States, whereas rice husks are popular in Southeast Asia, and animal husbandry residues (poultry litter) are popular in the United Kingdom.

Heat is the dominant mechanism to convert biomass into other chemical forms, for example, by torrefaction (pre-roasting), pyrolysis, and gasification, each of which depends on the extent to which the chemical reactions are allowed to proceed, and on the extent of available oxygen and process temperature. Other less common methods (experimental or proprietary) are hydrothermal upgrading (HTU) and hydro-processing.

The extent of power generated with biomass in the United States is about 11 GW of summer operating capacity, accounting for about 1.4% of the United States electricity supply. The largest biomass power plant in North America is the 140-MW New Hope Power Partnership facility, which uses sugar cane fiber and recycled urban wood as fuel to generate enough power for milling and refining operations. It also supplies renewable electricity to nearly 60 000 homes, and reduces the United States annual dependence on oil by more than 1 million barrels.

It's been said before that burning biomass is carbon neutral, and yet when it is used as a fuel it causes air pollution through emission of CO_2, nitrogen oxides (NO_x), volatile organic compounds (VOCs), particulates, and other pollutants, at times at levels above those from combustion of more traditional coal and natural gas fuels.

Black carbon – a pollutant produced from incomplete combustion of fossil fuels, biofuels, and biomass – has been claimed to be the possible second largest contributor to global warming. A Swedish study of 2009 on the giant brown haze that periodically covers large areas of South Asia found that biomass burning was the principal cause of the haze, and to a lesser extent fossil-fuel burning. The concentration of carbon-14 was significant and was associated with recent plant life rather than with fossil fuels.

In fact, the notion that biomass is carbon-neutral as was proposed in the early 1990s has been challenged by more recent studies which show that mature intact forests sequester carbon more effectively than areas cleared of trees. When a tree's carbon is released into the atmosphere in a single pulse, it contributes more to climate change than woodland timber rotting slowly over decades. Current studies also indicate that *even after 50 years, the forest has not recovered to its initial carbon storage*, and *the optimal strategy is likely to be protection of the standing forest.*

> *Forest bioenergy, as it is currently being developed in Canada, threatens the health of our forests and will harm the global climate for decades to come . . . The amount of wood being burned in power plants or turned into liquid fuels is growing exponentially without the public's knowledge and little government oversight or regulation.*

Nicolas Mainville, Greenpeace Canada

Recently, Greenpeace Canada and the United States Natural Resources Defense Council (USNRDC) have also questioned the notion that forest-based biomass has no impact on climate change (Figure 55). Recent scientific research also found that carbon released by burning biomass to be recaptured by re-growing trees can take many decades and even longer in low productivity areas. Moreover, logging operations disturb forest soils causing release of stored carbon. In light of the pressing need to reduce greenhouse gas emissions in the short term so as to mitigate current effects on climate change, a number of environmental groups oppose large-scale use of forest biomass to produce energy.

Figure 55 Cover of the magazine by Greenpeace (left) and picture of the forest biomass.

In their lengthy report titled *False Claims of Carbon Neutrality Conceal Climate Impacts*, Greenpeace Canada brought out some interesting points that also apply to the United States and elsewhere:

1) **Burning natural forest biomass** – whether for electricity, heat or biofuels – is *not* carbon-neutral as governments and companies claim. Burning trees contributes to climate change for decades, as shown by the most up-to-date science, until replacement trees fully grow back.

2) **Compared to current coal-fired electricity plants** in North America, current woody biomass power plants can emit at the smokestack up to 150% more climate-disrupting CO_2, 400% more lung-irritating carbon monoxide, and 200% more asthma-causing particulate matter to produce the same amount of energy. The CO_2 emitted will harm climate for decades before being captured by re-growing trees.

3) **The latest science** shows that burning biofuels derived from standing trees will emit more CO_2 emissions than using gasoline for well over a century.

4) **Burning boreal biomass** contributes to climate change through a long carbon payback time due to the slow re-growth of forests and the fragility of existing carbon stocks.

5) **Federal and provincial governments fail** to account properly the CO_2 emissions from forest bioenergy production by using the simplistic assumption of carbon neutrality. In truth, CO_2 emissions from biomass burning – about 40 megatons annually in Canada – are roughly the equivalent of Canada's 2009 light-duty vehicles emissions.

Renewable Energy – Biofuels

Biofuels were used in the United States at the beginning of the 1900s to power such cars as the Ford T model using ethanol as the fuel, after which interest in biofuels declined until Americans were confronted with the first oil crisis of 1973 and later the second oil crisis of 1979. Soon after the 1973 crisis, the Department of Energy established the National Renewable Energy Laboratory (NREL) in 1974 in Golden, Colorado. It began work in 1977 on such other things as research into biofuels. Two Energy Policy Acts voted by Congress, the first in 1994 and the second in 2005, promote renewable fuels.

The United States produces biodiesel and ethanol fuel (about 4.9 billion US gallons of ethanol; 18.38 million cubic meters) from corn as the main feed-stock – since 2005 it has overtaken Brazil as the world's largest ethanol producer. These two countries account for nearly 70% of all ethanol production, with total world production of 13.5 billion US gallons (51 million cubic meters, or 40 million metric tons). In 2007, the United States and Brazil were responsible for 88% of the 13.1 billion US gallons (50 million cubic meters) of total world production of ethanol alone. Increasing pollution control, climate change requirements, and tax relief increased expectations that the United States market for biofuels would continue to grow.

The largest biodiesel consumer in the United States is the Army. Most light vehicles on United States roads today run on blends of up to 10% ethanol/90% gasoline, although manufacturers already produce cars designed to run on blends with much higher ethanol content. One of the reasons for the popularity and demand for bioethanol fuel in the United States was the discovery in the late 1990s that the MBTE (methyl tertiary–butyl ether) oxygenate additive in gasoline was contaminating groundwaters. Current research focuses on cellulosic biofuels so as to avoid upward pressure on food prices and land use changes if use of biofuels were to experience a major growth.

Biofuels can also be obtained from the gasification of biomass; a growing number of people, although small at present, use wood gas to fuel cars and trucks across the United States. Biofuels are most popular in the farm states, where most of the biofuel feedstock is produced. The task now is to expand the market elsewhere. The transition from gasoline to biofuels is being driven by the coming of flex-fuel vehicles, as these allow drivers to choose different fuels based on price and availability. Corn ethanol and soybean biodiesel satisfy only 12% of gasoline demand and 6% of diesel demand.

The growth of biofuel industries has provided thousands of jobs in plant construction, operations, and maintenance, mostly in rural communities. The ethanol industry alone created nearly 154 000 jobs in the United States in 2005, boosted United States household income by $5.7 billion, and contributed about $3.5 billion in tax revenues to local, state, and federal governments. It's important to note that in 2007 the industry received $3.25 billion in federal support, not to mention other additional state and local support.

13
Energy UK

Adapt or perish, now as ever, is nature's inexorable imperative.

H.G. Wells

A direct consequence of the recent economic and financial crisis in Europe and elsewhere, but felt particularly by Europeans, has been a significant decrease in energy consumption and primary energy production – and not least a reduction of the trade deficit as a result of decreased energy imports. The United Kingdom is no different. Historically, the UK emphasized coal, nuclear, and natural gas as its principal energy mix, but now it appears that it will become a net energy importer. The total energy consumed in 2011 amounted to about 1.6% (that is, about $8.30\,EJ = 8.30 \times 10^{18}\,J$) of the estimated world's total of about $510\,EJ$ though the UK's population is about 1% of the total.

Primary Energy Resources

Figure 56 illustrates the energy available in the UK for the 5-year period between 2004 and 2009. Quite clearly, the quantity of primary energy decreased together with energy production as did electricity generated and emissions of the greenhouse gas CO_2. Only imports of energy grew during this time, at least up to 2008, but then tended to level off. The population grew ever so slightly from about 60 to 61 million.

The percentage of primary energy derived from major sources in 2007 is displayed in Figure 57. The two major sources of the energy mix were crude oil (38%) and natural gas – also at about 38% – followed by coal (ca. 17%), nuclear at about 6% and renewables at less than 2%. To meet the energy challenge, the UK addressed the long-term energy needs of the country in a 2007 White Paper in which it strategized that the UK should (i) reduce its greenhouse gas emissions by 60% by 2050 with a goal to show real progress by 2020, (ii) maintain energy supplies, (iii) promote competition in its markets and sustain economic growth, while (iv) ensuring that every household in the UK has an adequate supply of energy for heating purposes.

In July 15, 2009, the British Government launched a low-carbon Transition Plan in which it aimed to get 30% of its energy needs from renewables and 40% of

Powering Planet Earth: Energy Solutions for the Future, First Edition. Nicola Armaroli, Vincenzo Balzani, and Nick Serpone.

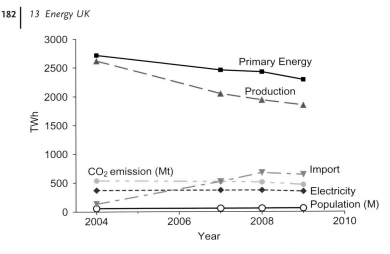

Figure 56 Energy in the United Kingdom for the period 2004 to 2009 in Terawatt-hours (1 Mtoe = 11.63 TWh). Population is given in million inhabitants and CO_2 emission is given in million tons. Data source: *IEA Key World Energy Statistics*.

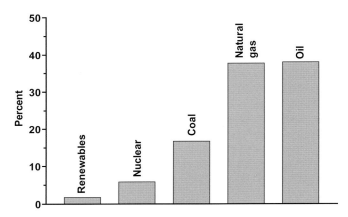

Figure 57 Percent of primary energy from major sources for 2007. Data source: *Digest of UK Energy Statistics 2007*. See: http://www.berr.gov.uk/energy/statistics/publications/dukes/page39771.html.

low-CO_2-content fuels to generate electricity by 2020. Specifically, the plan aimed at (i) protecting the public from immediate risk, (ii) preparing for the future, (iii) limiting the severity of future climate change through a new international climate agreement, (iv) building a low-carbon UK, and (v) supporting individuals, communities, and businesses to play their part.

To the extent that energy is an essential component of every aspect of people's lives and is crucial for a successful economy, the UK enacted an energy policy that includes production and distribution of electricity, fuel for transportation, and means for heating (preferentially through natural gas). The policy also spelled out two principal challenges, namely to reduce carbon dioxide emissions, and to ensure a secure, clean, and affordable energy supply as the UK becomes increasingly dependent on imported fuel. The government also engaged in other issues – the nuclear option, carbon capture and storage, renewables, and offshore gas and oil. The UK will need around 30–35 GW of new electricity generation capacity for the next two decades since the traditional methods with coal and nuclear power (built in the 1960s and 1970s) are no longer suitable as these two sources have reached their expected longevity and so are scheduled to be shut down.

While the energy policy is within the jurisdiction of the UK Government, the Scotland Act of 1998 gave Scotland the authority to enact its own energy policy, which is at variance with the UK's – it also has planning powers to put its policy priorities into effect.

Fossil Fuels

Total electricity production was 364.9 TWh in 2011, up about 18% from the 309.4 TWh generated in 1990 from the various sources displayed in Figure 58. In 1990, the major sources were coal, nuclear, and crude oil followed by imports and hydro – renewables were unknown in the UK at the time (0%). The picture emanating in 2011 tells a totally different story. Expectation in the early part of the 2000–2010 decade was that the total contribution from renewables to electricity production would rise to 10% by the end of the decade, whereas the Scots had targeted 17–18% of electricity generating capacity from renewables by 2010, rising to about 40% by 2020, a very optimistic prediction indeed.

The use of coal dropped more than 50%, while natural gas used to generate electricity jumped to nearly 40% from a nearly non-existent quantity of 0.05% in 1990 as a result of a combination of factors that led to the "dash for gas" in the 1990s. During this time the use of coal was severely curtailed owing (in part) to

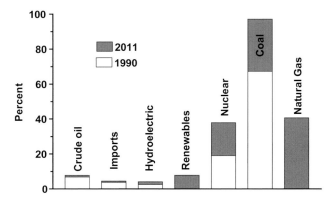

Figure 58 Percent electricity generated from the various sources in 2011 compared to the quantity produced in 1990. Data source: UK Department of Energy & Climate Change. See http://www.decc.gov.uk/assets/decc/11/stats/publications/dukes/5955-dukes-2012-chapter-5-electricity.pdf.

privatization of the coal industry, the introduction of laws that facilitated competition within the energy markets, and availability of cheap natural gas from the North Sea. One of the factors that led to the increase in natural gas use was the massive expansion of gas-fired generation capacity tied to the rapid construction of gas-fired power plants rather than coal-fired and nuclear power plants.

Nuclear remained essentially constant during this period, while new renewable energy sources began to contribute in the mid-1990s to the electricity generated, accounting for about 7.9% of electricity production in 2011. The use of hydro, imports and oil decreased in importance. Even though the North Sea oil began to flow in the mid-1970s, oil-fueled generation of electricity was relatively small then and decreased through the 1980s and 1990s up to 2011 (see Figure 58).

Driven by the need to produce weapons-grade plutonium in the early 1950s, the UK began to develop nuclear generating capacity. The first civilian nuclear power station – Calder Hall – began operation in August 1956 and was connected to the electricity grid; 26% of the UK's electricity was generated from nuclear power at its peak in 1997.

Fossil Fuels – Coal

In the nineteenth century during the industrial revolution, coal was the major energy source in the United Kingdom. In the 1940s, nearly 90% of the generating capacity came from coal, with oil providing the rest. Of late, however, coal has taken a back seat in favor of other forms of energy: natural gas and oil. Pressure to reduce sulfur (as SO_2) and carbon (CO_2) emissions has furthered the decline in the use of coal. Nonetheless, domestic coal remains an important energy source that provides the UK with security of energy supply, should the need ever arise. To this day, it still plays an active part in the UK's energy strategy due to its large domestic reserves, price stability, reduced capital expenditure, and time for the construction of a generating plant compared to nuclear power.

Between 1995 and 2009 the annual quantity of coal produced decreased from 37 million tons to about 7 million tons. Reserves in early 2009 stood at about 105 million tons – 45 million tons of which are easily accessible under current mining and investment conditions.

Fossil Fuels – Natural Gas

Domestic production of natural gas from the North Sea fields continues to decrease in spite of investments to enhance storage capacity of imported natural gas (mostly) from Norway as the UK is reluctant to place too much reliance on Russian natural gas supplies. Approximately 60% of natural gas consumed in the UK in 2011 was produced domestically, with the other 40% being made up by imports – yet the UK was self-sufficient just 6 years earlier in 2005. By 2021, North Sea oil and natural gas production are expected to fall 75% from 2005 levels, that is, to less than 1

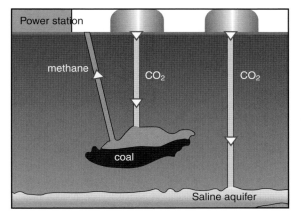

Figure 59 Planned technology to capture carbon emissions from a coal-powered electricity generating station. Source: http://news.bbc.co.uk/2/hi/uk_news/politics/6444373.stm.

million barrels annually. In this regard, Europe's oil (and coal) reserves are among the most tenuous in the industrialized world. Natural gas is likely to take a back seat in favor of other forms of energy as it looks like it will play a smaller part in the UK's future energy needs.

Over the past several decades, the UK has been fully aware of its past responsibility in greenhouse gas emissions ever since the start of the industrial revolution, and so together with most of the world's nations (170 of them) has committed itself to reduce emissions. In 2003, the UK was responsible for 4% of the world's greenhouse gases compared to 23% for the US and 20% for the rest of Europe. It expects to reduce its carbon emissions by about 60% by the year 2050, aided, in part, by trading carbon emission credits.

Between 1995 and 2004, the average carbon emissions from the transportation sector fell from 192 to 172 grams per mile. But since aviation fuel is not regulated by the Kyoto Protocol, the UK will still contribute to greenhouse gas emissions from its aviation sector. Nonetheless, the UK has been successful in reducing its carbon emissions from the road transportation sector. To further reduce its emissions from natural gas-fed power generation stations, it plans to use a technology – albeit in the planning stage at the moment – that involves carbon capture by seawater (Figure 59).

Nuclear Power

As we saw earlier, the first commercial nuclear power reactor in the United Kingdom began operating in 1956 – at its peak in 1997, 26% of the nation's electricity was generated from nuclear power. The share of nuclear power in electricity generation dropped to about 19% in 2004 and to 18% in 2011 resulting from the closure of some of its nuclear reactors. As of 2011, the UK operates 16 nuclear reactors at 10 plants, seven of which use advanced gas-cooled reactors (AGR), two use Magnox-type reactors, and one uses a pressurized water reactor. The UK also operates the nuclear reprocessing plant at Sellafield built in the late 1940s, with initial fuel loading into the Windscale Piles commencing in July 1950. Unlike the early United States nuclear reactors at Hanford in Washington State, which used a water-cooled graphite core, the Windscale Piles used an air-cooled graphite core.

The two Magnox plants and two of the seven AGR plants are to be closed by 2016, which will significantly increase the UK's energy gap. To offset this increase, some older AGR reactors have had their lives extended by 10 years. A report from the nuclear industry in 2005 raised concerns – as did earlier (2000) the Royal Commission on Environmental Pollution – that the UK could face a 20% shortfall in electricity generation capacity by 2015 unless the government took action to plug the energy gap.

One of the first moves in this direction was the announcement in 2006 of the construction of a conventional gas-fired power station. Then, proposals made in 2007 involved the construction of two new coal-fired power stations that were to be the first coal-fired stations to be built in 20 years.

Overall, the nuclear story in the UK has been debated at some length in the past decade beginning with the Government's 2003 Energy White Paper: *Our Energy Future – Creating a Low Carbon Economy*, which concluded:

> *Nuclear power is currently an important source of carbon-free electricity. However, its current economics make it an unattractive option for new, carbon-free generating capacity and there are also important issues of nuclear waste to be resolved. These issues include our legacy waste and continued waste arising from other sources. This white paper does not contain specific proposals for building new nuclear power stations. However we do not rule out the possibility that at some point in the future new nuclear build might be necessary if we are to meet our carbon targets.*

In April 2005, the then Prime Minister Tony Blair was advised that the UK should construct new nuclear power stations to meet the country's targets on reducing gas emissions responsible for climate change. Soon thereafter, January 2006, an energy review was launched to examine *the UK's progress against the medium and long-term Energy White Paper goals and the options for further steps to achieve them.* Critics took the review as a means to justify the building of new generation nuclear reactors – they further noted that the length of time needed to plan, construct, and commission nuclear power plants would not meet the shortfall in electricity generation and plug the expected Energy Gap.

On request by Greenpeace, in early 2007 the UK's High Court rejected the 2006 Energy Review because it was *seriously flawed* and it also held that the review's wording on nuclear waste disposal was *not merely inadequate but also misleading.* Subsequently, the government began new consultations, as it remained convinced that new nuclear power plants were essential to combat climate change and to minimize the UK's reliance on imported oil and gas. However, Greenpeace UK stuck to its view that carbon emissions could be cut by investing in a decentralized energy system that would maximize use of combined heat and power, and in renewable energy sources. That nuclear industry lobbyists could be connected to the governing Labor Party didn't go unnoticed in the media.

Further consultations led to the 2007 Energy White Paper which contained the government's preliminary view that the private sector be allowed to invest in new nuclear power stations. A report by a consulting firm further suggested that preference be given to locate new power stations on existing nuclear stations sites owned either by the Nuclear Decommissioning Authority or by British Energy. Greenpeace UK held to its earlier views that new reactors built on old sites would only reduce the UK's total carbon emissions by 4%. Was this a sufficient reason to build new reactors?

In a speech to Greenpeace UK in December 2007, the current Prime Minister, David Cameron, suggested replacing the large-scale electricity generation stations owned by the government and big energy companies with a *decentralized energy* system of Combined Heat and Power (CHP; also known as cogeneration), which involves the use of a heat engine or a power station to generate simultaneously electricity and useful heat. No mention was made of nuclear power.

In January 2008, the UK Government gave its go-ahead for the construction of new generation nuclear power stations. Scotland opposed such a move on its soil, countering that focusing on new nuclear plants would undermine efforts in search of cleaner, greener, more sustainable, and secure energy sources. Moreover, it decided to concentrate on renewable energy sources and on greater energy conservation.

By November 2009, the UK Government had identified 10 nuclear sites that could accommodate the new nuclear reactors, two of which were new sites. Later (October 2010), however, two of the sites were ruled out. Cost estimates of replacing Britain's ten nuclear power stations put the price tag as high as £48 billion (UK pounds), excluding the costs of decommissioning the old reactors and treating and storing nuclear waste.

In an article that appeared on 21 February, 2010 in *The Times*, London, Jonathan Leake reported that the Research Councils UK had committed to investing in a 20-year study and construction plan of a ***nuclear fusion*** power station to commence operating around 2030. The Councils questioned the article's accuracy.

The 2011 Fukushima incident delayed the UK Government's program to build new nuclear power stations located in England by at least 3 months until lessons were learned from the Japanese accident. The Government was criticized for collusion with three major nuclear reactor companies in manipulating communications so as to maintain public support for nuclear power.

What the future of nuclear power stations will be in the UK is not clear. Current policy favors construction of new nuclear power stations that will be left to and financed by the private sector, though some government participation and long-term liabilities will remain, albeit in a limited way. The private sector is interested in taking over the building and operation of new nuclear power stations, but only under certain conditions: (i) that the government set a suitable carbon price on coal and gas electricity generation, and (ii) that the operators be provided with government incentives such as capacity payments.

Third-generation nuclear reactors have been said to offer some improvements from earlier designs: (i) simpler designs with less materials and less on-site fabrication, (ii) standardized designs that can cut down costs, (iii) improved project management, (iv) competitive contracts, (v) turn-key (fixed costs) contracts rather than cost-plus ones, and (vi) recent evidence from China and South Korea that reactors can be built faster and at lower cost.

Nuclear Waste Management and Disposal

A thorny issue that has not been resolved in the UK, as has been the case elsewhere where nuclear reactors are an important component of a country's energy mix, is the management and disposal of nuclear wastes. The UK possesses radioactive wastes from its nuclear weapons program and from its nuclear power stations. So far, the management of these wastes has been the Government's responsibility, and though new nuclear power stations will be the responsibility of the private sector, it appears that nuclear wastes from all sources will be stored at Sellafield, an off-shoot from the original Windscale nuclear reactor site, where the reactors are currently undergoing decommissioning and dismantling (see below). A neighbor of Windscale, Calder Hall, is also undergoing decommissioning and dismantling of its 4 nuclear power-generating reactors.

In 2006, the Committee on Radioactive Waste Management recommended that nuclear wastes be disposed of geologically by burial at depths between 200 and 1000 meters with no future retrieval. Luckily, implementation of this recommendation is not likely to take place for several decades to come, as there are some additional thorny issues yet to be resolved, such as the social and ethical concerns of the UK people. The report was not without its critics – David Ball, professor of risk management at Middlesex University, who had resigned from the Committee in 2005, stated that the report was based more on opinions than on sound scientific grounds.

A further report by the Committee in 2008 indicated that selection of a suitable site should rely on volunteerism with the surrounding region being rewarded infrastructure investment, job creation for the long term, and an additional tailored interesting package. Volunteerism or Briberism? Really!

Windscale Fire and Decommissioning

Despite its highly skilled nuclear labor force and its nuclear technology, the UK has not been impervious to nuclear accidents. On October 1957, a fire in Pile 1 (Figure 60) in the Sellafield complex destroyed the nuclear core and released some 250 Terabecquerels, that is, 20 000 curies of radioactive material (for example iodine-131) to the surroundings. Pile 1 was then no longer serviceable, and though Pile 2 was not damaged by the fire, it was also shut down as a precautionary

Figure 60 Photograph of the Windscale Piles being decommissioned (Chris Eaton, October 2009). Source: http://www.geograph.org.uk/photo/330062.

measure. This did not affect the UK's nuclear weapons program as it had enough plutonium for some atomic bombs.

Decommissioning, dismantling, and cleaning up of both Piles began in the 1990s under the supervision of the United Kingdom Atomic Energy Authority, but by 2004, Pile 1 still contained 15 tonnes of uranium fuel. Complete decommissioning is not expected until at least 2037, at some significant cost to taxpayers. The 2002 estimates to decommission the Sellafield site set the cost at £48 billion, which 5 years later was raised to £73 billion, which included the cost of continued operation of the reactors for their remaining life. In May 2008, the figure of £73 billion was expected to increase by several billion pounds.

While governance of a country and management of a large national facility should not be based on opinion polls in a situation that involves the use of nuclear power to generate electricity, political leaders would be wise to consult the populace, particularly those in the proximity of nuclear power plants. Some studies found the presence of clusters of leukemia cases in unborn infants in the vicinity of some nuclear plants, although similar clusters were also discovered in areas distant from nuclear plants. Another study carried out in 2003 found no evidence of increased childhood cancer around nuclear power plants, but did find an increased number of leukemia and non-Hodgkin's lymphoma (NHL) cases near other nuclear installations including the Sellafield complex. Such findings could not be reasonably explained, but they were deemed not to be due to sheer chance.

A November 2005 poll, conducted for the well-respected business advisory firm Deloitte, found that 36% of the UK population supported the use of nuclear power, though 62% would support an energy policy that combined nuclear along with renewable technologies – 35% of the respondents had expressed a wish that much of the UK's electricity demand come from renewable sources within 15 years – that is, by 2020 – which was double the expectation of the UK government.

A subsequent 2010 British survey found public opinion divided on energy issues, with the majority expressing some concerns on nuclear power and low confidence in the government and the nuclear industry. Clear preference was expressed for renewable energy. A more recent poll (2011), taken after the Fukushima incident, saw the support for nuclear power decrease by nearly 12%, while other polls appeared to indicate otherwise – that is, increased support for the nuclear option.

Apparently, in the latter poll conducted for the British Science Association, over 40% of respondents said the benefits of nuclear outweighed the risks – respondents believed that future energy security was more important than nuclear risks. This bucked the trend observed in Italy, Germany, and most of all in Japan, that lived through Fukushima.

Reversing the British trend, Scottish leaders have made it clear that nuclear power was not an option, even though at present Scotland derives nearly 50% of its electricity from two nuclear power plants. It hopes to replace these plants with renewables when the reactors cease operation – one in 2016 and the other in 2023.

Renewable Energy

Historically, renewable energy has not been a major player in the energy mix of the United Kingdom, although it had the potential from wind power and tidal power. Only since the mid 1990s did renewable sources begin to contribute to electricity production, albeit only to a small extent. Hydroelectricity has not been a viable option for the UK since its rivers lack the necessary force.

Government energy reviews of the last decade have set various targets for electricity generation from renewable energies. For instance, the 2002 Energy Review established that 10% of electricity was to be produced from these sources by 2010/2011; the target was later increased to 15% by 2015. The 2006 Energy Review set 20% as the target for 2020. For Scotland, the goals were 17–18% electricity from renewables by 2010, to increase (optimistically) to 80% by 2020.

By 2004, the amount of electricity generation from renewables was 250 MW – it increased to 500 MW in 2005. According to statistics released by the UK's Department of Energy and Climate Change, in 2009 renewable energy sources provided 6.7% of electricity domestically generated and 9.6% in the second quarter of 2011. This represents 7.86 TWh, which met the target established back in 2002.

Statistics also confirmed Scotland's leading role in renewable electricity. In 2010, it had around 20% more generating capacity from renewables than England, though generation of electricity in England was 45% higher owing to its intensive use of biofuels from domestically grown sugar beet.

Renewable Energy – Wind Power

At the end of 2011, the installed capacity of wind power in the UK was 6540 MW, placing the UK as the world's 8th largest producer. Wind power is the second largest source of renewable energy after biomass. The British Wind Energy Association, now known as RenewableUK, estimated in 2010 that more than 2 GW of capacity could be deployed annually by wind power for the years 2011–2016. But let's back-track for a moment.

The January 2009 European Union Renewables Directive put a target of 20% of the EU's supply of *final* energy from renewable sources by 2020. The UK's allocated target was 15%. As renewable heat and fuel production in the UK were at extremely low bases, RenewableUK estimated that this would require 35–40% of the UK's electricity to be generated from renewable sources by that date, to be met largely by 33–35 GW of installed wind capacity. At the end of 2007 the UK had already planned to expand wind energy up to 25 GW of wind farm in offshore sites in preparation for a new round of development. This was in addition to the 8 GW planned in 2001 and 2003. Taken together, it was estimated that this would result in the construction of over 7000 offshore wind turbines.

Some Historical Notes

The first wind farms in the UK were built onshore, and currently generate more power than offshore farms. The first commercial wind farm, built in 1991 in Cornwall, consisted of 10 turbines, each with a capacity to generate a maximum of 400 kW. The early 1990s saw a small but steady growth, with about 6 farms becoming operational each year. The larger wind farms tended to be built on the hills of Wales. Smaller farms started to appear on the hills and moors of Northern Ireland and England. The first commercial wind farm in Scotland began operation at the end of 1995. The late 1990s saw sustained growth, and in 2000 the first 1-MW turbines were installed. The pace of growth started to accelerate as the utility companies Scottish Power and Scottish & Southern became increasingly involved in meeting legal requirements to generate a certain amount of electricity from renewable sources.

Wind turbine development continued rapidly, and by the middle of the last decade (2000–2010) the norm in turbine deployment became the 2-MW+ turbines. Growth continued with bigger farms that employed larger, more efficient turbines sitting on taller and taller masts. The UK's first 100-MW+ farm went operational in 2006, which also saw the first use of a 3-MW turbine.

The largest onshore wind farm in England was completed in 2008. The repowering of another wind farm created the largest farm in Northern Ireland. The largest wind farm (140 turbines) in the UK went live in 2009 at Whitelee on Scotland's Eaglesham Moor, generating 322 MW. That year, UK onshore wind farms generated 7564 GWh, or a 2% contribution to the total UK electricity generation (378.5 TWh).

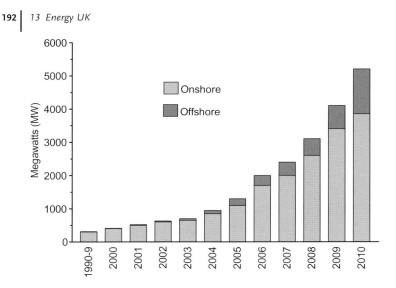

Figure 61 Growth of installed onshore and offshore wind power capacity in the United Kingdom for the period 1990–2010. Source: http://en.wikipedia.org/wiki/File:UK_windfarm_growth.PNG.

Three offshore wind farms came on stream in 2010. Over 1100 MW of new wind power capacity was brought online during 2010, a 3% increase from 2009. There was a 38% drop in onshore installations to 503 MW compared with 815 MW in 2009, but there was a 230% increase in offshore installations with 653 MW installed (285 MW in 2009). Figure 61 illustrates the electricity generated over the 1990–2010 decades from wind power from onshore and offshore wind farms. Several more 100-MW+ wind farms are scheduled to be built on the hills in Scotland – they will be provided with the new 3.6-MW turbines.

As of December 2011, there were 321 operational wind farms in the UK, with 3506 turbines and 6540 MW of installed capacity. Over 3500 MW of wind farms are currently under construction, while another 5590 MW have planning consent and some 10 000 MW are in the planning stage awaiting approval.

Large onshore wind farms tend to be connected directly to the National Grid, whereas the smaller wind farms are connected to a regional distribution network, termed *embedded generation*. In 2009, nearly half of wind generation capacity was embedded generation, expected to decrease in the future as larger wind farms are built.

Figure 62 displays a photograph of the skyline of the 24-MW Ardrossan wind farm in North Ayrshire, Scotland – it was officially opened on August 10, 2004.

The well-respected newspaper, The Guardian (London), reported that the Ardrossan Wind Farm had been *overwhelmingly accepted by local people*. Instead of spoiling the landscape, locals believe it has enhanced the area: *The turbines are impressive looking, bring a calming effect to the town and, contrary to the belief that they would be noisy, we have found them to be silent workhorses.* Unfortunately, one of the turbines of the wind farm failed catastrophically in a ball of fire during the severe storms of December 2011.

Figure 62   Skyline of the Ardrossan wind farm, North Ayrshire, Scotland. Note the houses in front. Photo by Vincent van Zeijst, July 3, 2010. Source http://en.wikipedia.org/wiki/File:Ardrossan,_Scotland,_United_Kingdom.JPG.

Figure 63   Scroby Sands offshore wind farm, Great Yarmouth, Norfolk, UK, seen from the beach (photo by Anke Hueper, Karlsruhe Germany; November 04, 2005). See http://en.wikipedia.org/wiki/File:Scrobysands04.11.2005.a.jpg.

Figure 63 shows a photograph of the Scroby Sands offshore wind farm, Great Yarmouth, Norfolk, UK, as seen from the beach.

The United Kingdom became the world leader of offshore wind power generation in October 2008 when it overtook Denmark. It also has two of the largest offshore wind farms in the world, the Thanet wind farm (commissioned in 2010), located off the Kent coast, and the recently (2012) commissioned Walney offshore wind farm (see Table 9, Chapter 8). Currently it has 1525 MW of operational nameplate capacity (i.e. technical full-load sustained output), with a further 2054 MW in construction. The UK possesses over a third of Europe's total offshore wind resource, equivalent to 3 times the electricity needs of the country at current rates of electricity consumption, although this is only at certain times. It has been estimated that to meet EU targets, the UK will have to install 7500 additional offshore turbines by 2020.

Renewable Energy – Solar Power

In countries like Germany, installation of solar electricity receives substantial Government support, stemming from their plans to phase out nuclear energy. Germany subsidizes solar electricity to such an extent that by 2006 it had already installed 3.0 GWp (GigaWatt-peak), representing nearly 90% of all European capacity of 3.4 GWp. For comparison, at the end of 2006 the UK's installed photovoltaic capacity was about 13 MWp – only about 0.3% of the European total.

Because of the geographical location of the UK in Europe (Figure 64), solar power has been a minor source of renewable energy in that country, where insola-

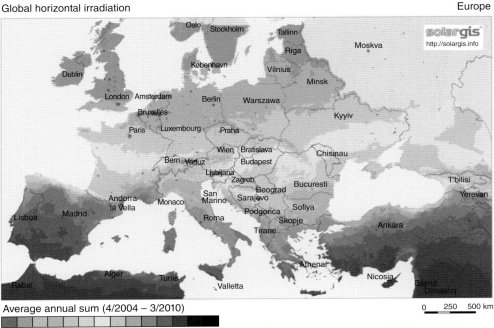

Figure 64 Global horizontal solar irradiation. Source: Copyright 2011 Geo-Model Solar s.r.o. See: http://upload.wikimedia.org/wikipedia/commons/5/59/SolarGIS-Solar-map-Europe-en.png.

tion is less than 120 W/m² (that is, 2.9 kWh/m²/day, or 1050 kWh/m²/year). This is only a fraction of what is available in subtropical zones such as southern Spain and North Africa. It should be noted that the higher wind speeds prevalent in the UK can cool photovoltaic (PV) modules, leading to higher efficiencies than would be otherwise expected at these levels of insolation. As of August 2011, about 875 MW of solar photovoltaic power was installed in the UK capable of producing annually about 900 GWh of electricity.

Environmentalist George Monbiot advocates replacing fossil fuels with carbon-free energy sources, but says that subsidies for solar power is a *terrible investment* for the UK, apparently forgetting that subsidies have been very important in the development of renewables in other countries such as Germany, Italy, and Spain.

On the other hand, Derry Newman, chief executive of Solarcentury, has argued that the UK's famously overcast weather does not make it an unsuitable place for solar power, as solar panels work on daylight, not necessarily direct sunlight, forgetting to say that the efficiency is much smaller.

Construction of the largest solar park in the United Kingdom was completed in July 2011 just 7 weeks after being granted planning permission. The 5-MW free-field system, located in the parish of Hawton near Newark-on-Trent in Nottinghamshire (Figure 65), will feed back 4860 MWh of electricity to the national grid. There are several other examples of 4–5 MW field arrays of photovoltaics in the UK. It is unlikely that such large arrays will be built in the future beyond

Figure 65 Conergy's 5-MW free-field system located in the parish of Hawton near Newark-on-Trent in Nottinghamshire is the largest in the UK. Photo: Lark Energy. Source: http://www.solarpowerportal.co.uk/news/conergy_completes_record-breaking_uk_solar_park/.

August 2011 as cuts to the feed in tariff, announced by the Department of Energy and Climate Change in June 2011, have made large solar photovoltaic arrays (greater than 250 kW) far less attractive to invest in by interested developers.

Installation of a residential photovoltaic system for an average-sized house in the UK can cost around £5000–£8000 per kWp installed – most residential usage of solar electricity requires between 1.5 and 3 kWp. Once installed, such systems can yield annual savings between £150 and £200.

Renewable Energy – Geothermal Energy

The 1973 oil crisis prompted the UK to investigate and exploit possible geothermal power sources, but the efforts were soon abandoned as the price of oil fell soon thereafter. According to the British Geological Survey, the most favorable (low enthalpy) geothermal energy source in the UK is the Permo-Triassic sandstones that extend deep into depositional basins and in many cases onshore extensions of major offshore basins. The basins of principal interest are located in East Yorkshire and Lincolnshire, Wessex, Worcester, Cheshire, West Lancashire, Carlisle, and basins in Northern Ireland.

In the 1980s, the UK's Department of Energy undertook a research and development program to further examine the potential of geothermal aquifers. After some initial success, the well drilled in 1981 in the Wessex Basin was deemed too small to be commercially viable. The project was abandoned. However, the Southampton City Council decided to create the UK's first geothermal power scheme as part of a plan to become a self-sustained city in energy generation. Eventually, the scheme was developed, with construction starting in 1987 on a well that would draw water from the Wessex Basin aquifer at a depth of 1800 meters and at a temperature of 76 °C. Hot brine from the geothermal well provides 18% of the total district heating mix, with fuel oil accounting for 10% and natural gas for 70% of the remainder. Southampton claims to be the greenest city in the UK.

According to its own website, after ten years of operation the scheme in Southampton delivers annually more than 30 000 MWh of heat alongside 4000 MWh of electricity sold from the generating plant (Figure 66) plus 1200 MWh of power providing chilled water on tap. It saves over 10 000 tonnes per year of CO_2 emissions in the process. The scheme serves 20 major consumers in the city center. Circulating water is pumped around the city through 11 km of insulated service pipes within a 2 km radius of the heat station with just 0.5 °C/km temperature loss.

In 2004, a scheme was announced that would heat the UK's first geothermal energy model village near Eastgate, County Durham. To this effect, a planning application was submitted in 2008 for a hot rocks project on the site of a former cement works. The project was to use the Hot-Dry-Rock geothermal technology to heat water pumped below ground onto geothermally heated rock.

Another area with great potential for geothermal energy is the continental shelf in the North Sea. At the moment, hydrocarbons are extracted from this region,

Figure 66 Photo showing the heat station in Southampton. Source: http://ec.europa.eu/ energy/res/publications/doc2/EN/SOUTH_EN.PDF.

with each year seeing the output fall by 5%; in the not too distant future, however, fossil fuel extraction will become uneconomical. An alternative use could be geo-thermal power generation. In this regard, a 1986 study showed that the continental shelf in the UK, where the North Sea platforms are located, has a relatively thin earth's crust, giving the wells high bottom hole temperatures. Heat from these wells could therefore be utilized to generate electricity and, by the use of submersible cables, help power the national grid.

The Eden Project in Cornwall was given permission in December 2010 to build a Hot Rock Geothermal Plant, with drilling commenced in 2011 and electricity to be produced from the second half of 2013. The plant, located on the north side of the Eden Project, is a showcase of environmental projects at Bodelva, near St Austell. It is expected to produce up to 4 MW of electricity for use by Eden with the surplus large enough for about 5000 houses to go to the National Grid.

Renewable Energy – Wave and Tidal Power

The geographical location of the United Kingdom makes it most suitable to exploit its great potential of generating electricity from wave power and tidal power. To date, however, wave and tidal power have received very little attention for development and consequently have not yet been exploited on a significant commercial level due to doubts over their economic viability. By contrast, in February 2007, Scotland announced funding for the UK's first wave farm, which when built will be the world's largest, with a capacity of 3 MW at a cost of over 4 million pounds.

Renewable Energy – Biofuels

Biogas has already been exploited in some areas and was the UK's leading renewable energy source, representing 39.4% of all renewable energy produced (including hydro). In 2004 it provided 129.3 GWh of electricity, an increase of 690% from 1990 levels. Other biofuels could provide a close-to-carbon-neutral energy source, if locally grown. However, experience in South America and Asia has shown that production of biofuels for export has, in some cases, resulted in significant ecological damage.

Electricity in the United Kingdom

For its electricity requirements, the UK relies principally on fossil fuels and between 15 and 20% on nuclear power. In the five years between 2004 and 2009, electricity consumption dropped by 11% (per capita 736 kWh), while the renewable energy share of total electricity use increased 2.8%. Wind power share was 3.2% of electricity compared to top countries like Denmark (24% of electricity), Spain (14.4%), Portugal (14%), Ireland (10.1%) and Germany (9.3%) for this same 5-year period. Use of light bulbs in the UK ended voluntarily in 2011.

In 2004, gross production of electricity was 393 TWh, placing the UK in the 9th position of the world's top producers. Natural gas was responsible for 160 TWh in 2004 and for 177 TWh in 2008. The percent share of electricity generated from renewable energy sources for 2009 in the United Kingdom is shown in Figure 67, while the quantity produced in GWh is shown in Figure 68. Clearly, wind power contributes the greatest share, followed by biomass, hydro, and biogas.

In terms of gigawatt-hours of electricity produced, onshore wind power contributes the most out of all the renewables. Wave and tidal power, together with solar photovoltaics, contributed a negligible amount.

In summary, as stated in its 2007 White Paper, the UK Government's proposed strategy for its energy outlook is based on a number of practical measures, namely:

1) establish an international framework to tackle climate change, including stabilization of atmospheric greenhouse gas concentrations and a stronger E.U. Emissions Trading Scheme;

2) provide legally binding carbon targets for the whole UK economy, reducing emissions through the implementation of the Climate Change Bill;

3) make further progress in achieving fully competitive and transparent international markets, including further liberalization of the E.U. energy market;

4) encourage more energy saving through better information, incentives and regulation; and

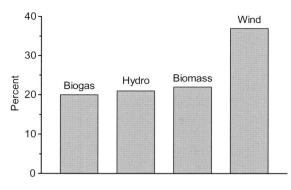

Figure 67 Percent of electricity generated from renewable energy sources for 2009 in the United Kingdom. Data source: Department of Energy and Climate Change: Digest of United Kingdom energy statistics (DUKES), p. 184. See http://www.decc.gov.uk/assets/decc/Statistics/publications/dukes/313-dukes-2010-ch7.pdf.

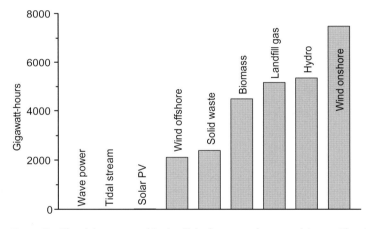

Figure 68 Electricity generated in the United Kingdom from renewable energy in 2009. Data source: Government report of Renewables Growth to 2020, See http://www.decc.gov.uk/assets/decc/11/meeting-energy-demand/renewable-energy/2185-analysis-of-renewables-growth-to-2020-aea-report.pdf.

5) provide more support for low-carbon technologies, including increased international and domestic public–private sector collaboration in the areas of research, development, demonstration, and deployment, for example, though the launch of the Energy Technologies Institute and the Environmental Transformation Fund.

14
Global Trends

Even some of the greatest technology-led revolutions, or allegedly technology-led, really were only made possible because of trends already present.

Scott Cook

In the last four chapters we have examined the energy resources in Italy, Canada, the USA, and the UK in terms of energy availability and energy consumption, and what is in store for these countries as we move forward. In this regard, if the objectives indicated for Italy appear unrealistic, we should mention that other countries have even more specific ambitious projects. In 2005, Germany was at an almost identical level to Italy's with regard to renewable energy at 5.8%, and has a 2020 target similar to Italy's (18%). However, Germany is running at more than twice the speed of Italy – Germany will likely have reached 30% by 2020.

Accordingly, it seems rather obvious that Italy is likely to buy renewable energy by 2020 if it doesn't have the political will to respect its commitments at the European level. In addition to household appliances and cars, German companies will also supply Italy with the energy to get Italians to work – a nice all-inclusive service. In the meantime, Germany is already working hard toward the production of 80% energy from renewable sources by 2050 – a long-term strategy aimed at the dominance of its industries for many decades in the latter half of the twenty-first century.

Global growth data concerning the two most promising renewable technologies are impressive. Wind power installed from 2000 to 2011 is more than tenfold, having reached the 239 GW threshold. Today, wind energy produces 512 TWh per year globally, that is, 3% of the world's electricity production, equivalent to 75 nuclear plants or coal-fed 1000-MW power installations operating continuously.

The photovoltaic (PV) option started a decade late and, as Figure 69 demonstrates, can count on producing about 68 GWp power – that is, 80 TWh/year, equivalent to twelve 1000-MW conventional power stations.

More than anything else, renewable energy technologies produce electricity. This will give further impetus to a process that has been in place for some time: the energy system will be based increasingly on electricity. This phenomenon is also extending slowly to the transport sector, which is the most established and therefore the least prone to change. Not by chance and for the first time, in 2011 the car of the year award went to an electric car: the Nissan Leaf model.

Powering Planet Earth: Energy Solutions for the Future, First Edition. Nicola Armaroli, Vincenzo Balzani, and Nick Serpone.
© 2013 Wiley-VCH Verlag GmbH & Co. KGaA. Published 2013 by Wiley-VCH Verlag GmbH & Co. KGaA.

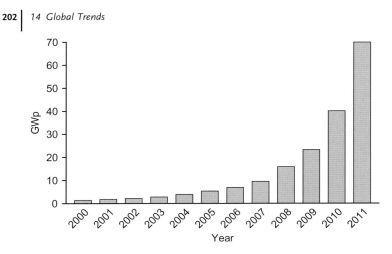

Figure 69 The evolution of photovoltaic power installed worldwide in the period 2000–2011. Source: EPIA Report 2011.

The increased electrification of our civilization will require decisive technological advances in storage systems to solve the problem of intermittency of primary energy sources, first and foremost the Sun and wind. There will be a need to develop batteries, capacitors, superconductors, underground storage of compressed air, electrolysis of water to produce hydrogen, and many other systems. Even the current electrical grids (networks) will have to undergo a profound transformation. The current grids, designed 100 years ago, were created to transmit an electric flow produced by a few large power plants, whereas tomorrow's grids, or so-called intelligent networks (*smart grids*), will guarantee stability in a system consisting of millions of small-scale producers and accumulation systems, whether fixed or mobile, exchanging energy between each other in the same manner as people currently exchange information via the Internet.

Our descendants will be smiling at our concerns about the intermittency of renewable sources. The network can always rely on the fact that half the planet is constantly illuminated by the Sun at any time. Not even the seasonal oscillations will be a problem, because when the northern hemisphere is in the winter season, the southern hemisphere is in the summer season, and *vice versa*.

A Shot at the Wrong Target

The refusal of the nuclear option by Italians, reiterated in the 2011 referendum, has prompted a wailing campaign mocking renewable energy sources and technologies. Some commentators, who had previously talked about the bright future of nuclear power in Italy and worldwide, have suddenly become experts in demonstrating that Italy's energy future rests with coal. Renewable energy sources, they say, are too costly and basically irrelevant – in total contrast to what's happening elsewhere. To refute such incredible beliefs, let's look at some key data on renewable energy, starting with costs.

Incentives or subsidies for the production of electricity from renewable sources in Italy amounted to 2.7 billion Euros during 2010. This is less than half the *system charges* that Italians pay annually on their utility bills (5.5 billion Euros), which include, among other items, the cost of dismantling old nuclear facilities (400 million Euros in 2010 alone) and the notorious *sources assimilated to renewables* or *CIP6* (1.2 billion Euros). This is nothing short of an immoral rip-off of Italian customers, who have been paying for the last 20 years – a total of some tens of billions of Euros – to promote electricity production from such sources as, for example, wastes from oil refineries and waste incinerators that have nothing to do with renewable energy sources.

A study led by the Polytechnic Institute of Milan estimated that over 60% of the 2.7 billion Euros incentive for renewables, found its way from citizens' and businesses' pockets – through energy consumption bills – directly into Italy's Treasury coffers by means of fiscal mechanisms. To this must then be added the fact that renewable sources employ tens of thousands of people, thereby generating more wealth for the State. Moreover, production of renewable energy should allow Italy to limit (if not avoid) European sanctions for breaches of the quotas to produce energy from renewable sources and to reduce CO_2 emissions to which it committed itself.

If we were now to account also for the negative externalities (discussed in Chapter 5) and for some savings that Italy could make by limiting its military actions, as it already happened in the case of Libya, a country that has assured its oil and gas supplies for years, then we could have a comprehensive picture of the situation.

To maintain that incentives/subsidies intended for the exploitation of renewable energy sources represent an unsustainable economic weight for the Italian economy is, in short, simply fraudulent. Perhaps there are special interests at play?

The plan for Italy briefly described earlier would have a predictable cost around 500 billion Euros over the next 10 years (that is, 50 billion Euros annually). One can easily infer that benefits far exceed costs. But if someone were consistently convinced that such cost was totally unsustainable by Italians and by Italy's economy, then we invite you – the reader – to give some thought to the following data.

Every year some 2 million new cars are sold in Italy. If we assume an average cost of 15 000 Euros to buy a car, the total would amount to some 30 billion Euros annually. Currently, Italians own 36 million cars. Considering a conservative average expenditure of about 3000 Euros per year in operating costs (for fuel, insurance, license plates, highway maintenance, and so on), Italians spend annually 108 billion Euros.

In summary, in using their cars, Italians spend around 140 billion Euros *every year*. Yet no one could ever convince Italians that this sum was money down the drain, as the car provides a convenient service which they have no intention of giving up, though a car weighs heavily on their household budget. The problem is therefore not an economic one, but one of mindset. When Italians become convinced – perhaps driven more by necessity than by reasoning – that they need to take charge of producing energy, not only collectively but also personally, then everything might become easier on their pocket book.

With good fortune, technological advances and the market may help. The production of electricity from photovoltaic panels and from wind power will have production costs lower than conventional technologies much earlier than 2020. Other renewable technologies may follow. Even the much hated incentives/subsidies will become history and the rooftop solar panels (whether thermal or photovoltaic) will be accessories that will be taken for granted, just as TV antennas and TV dishes are.

Sustainability of the Photovoltaic Option

Short of economic motivations, detractors of renewables often kick up a storm on the environmental and energy sustainability of photovoltaic panels.

The first classical objection – now obsolete but hard to eradicate – is that photovoltaic panels produce less energy than the energy spent to fabricate them. In fact, in energy terms today's payback time for these devices is between 1 and 3 years – compare this with their operating period of at least 25 years.

The second objection – an authentic urban myth – is that the panels generate highly – albeit unknown – toxic wastes. In fact, a silicon photovoltaic panel is made of non-toxic materials such as silicon (Si), aluminum (Al), copper (Cu), silver (Ag), with Si being highly abundant in nature (sand), while Al, Cu, and Ag are fully recyclable. The same is true for thin-film panels, even when they contain toxic elements such as cadmium (Cd).

To overcome these concerns, it's worth mentioning that there is an international consortium of PV industries (www.pvcycle.org) engaged in collecting and recycling all the panels currently used in the domestic and industrial sectors when the panels reach the end of their useful life. Clearly, a deep difference emerges in the vision and sense of responsibility. The PV industry is concerned about how it will treat wastes in 20–30 years, since the nuclear industry has not yet found a solution to the problem of treating radioactive wastes produced 60 years ago. From these details we can perceive where energy issues stood in the past and where energy issues will arise in the future.

Finally there is the thorny question of where to install the panels. In recent years, photovoltaic installations have proliferated on land. This has led to quite a few misgivings about neutralizing a significant quantity of land and altering the natural landscape. It must be said, however, that the land used for such installations had already been destined for industrial activities. Hence, the alternative to PV installations on land in Italy would have been the far more invasive concrete industrial warehouses, and in that case no one would have objected.

There is no doubt that the preferred location of new photovoltaic panels is on rooftops of existing residential, commercial, and industrial buildings. However, concerns for unsustainable consumption of land are unfounded. We noted in Chapter 7 that to satisfy *its entire electricity needs* Italy would require installing PV panels over a landmass of about 2400 square kilometers. If installed on land, these panels would cover only 3.3% of the arable land and 1.7% of Italy's total agricultural land. Actually, by 2020, only a tiny fraction of that surface would suffice to install photovoltaic panels. In fact as we have already noted, Italy needs an additional 16 GWp that could be obtained from just 120 square kilometers of landmass (assuming 7.5 square meters of panels per kWp), equivalent to 60 square meters of PV panels per rooftop on two million buildings.

Of course, at the local level, administrations must ensure that photovoltaic systems do not upset specific vocational agricultural landmasses, as the latter represent a priceless resource that must be preserved, even if Italy's agriculture is going through some difficult moments owing to recent negative aspects of the global economy.

Will Renewable Energy Sources Suffice?

As we shall describe in some detail next (Chapter 15), consumption of primary energies at an annual average of 2.6 tons of oil equivalent (toe) per person could afford and sustain a reasonable good quality of life. At present, the average global value *per capita* is substantially lower – about 1.8 toe per year (Table 5, Chapter 4). If in 2050 the world population were to reach 9 billion people, the estimated global energy demand – that is, energy to secure dignity to all the inhabitants of the planet – would be almost double what it is today at 24 000 Mtoe of current world energy consumption. Accordingly, the question that arises is whether the world will be able to meet the energy requirements at these levels relying solely on renewable energies. Even though there is wide consensus among scholars that the answer is yes, it is difficult to assess the actual potential of renewable energy that is technically exploitable.

Figure 70 compares the current daily global energy consumption with estimates of the potential of major renewable sources that are theoretically exploitable and with estimates of those sources that are already technically exploitable and environmentally sustainable using presently available technologies.

In practice, the world could already run with renewables even though many technologies are not yet mature. If this does not happen it's because the current system based on fossil fuels is too well entrenched. It would take decades to over-turn the mindset on these conventional fuels, although the exercise is far from impossible. Whoever argues that renewables can never replace uranium and fossil fuels is definitely not well informed.

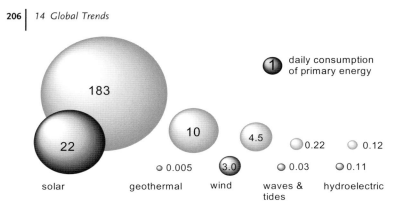

Figure 70 Total theoretical potential (light gray spheres) and presently technically exploitable (dark gray spheres) renewable energy sources. The black sphere is a reference representing the current world's total daily energy consumption set equal to 1.

But There Is Always a Limit

Two prominent American scientists, Mark Jacobson and Mark Delucchi, recently published a proposal to produce 100% of the world's energy demand through renewable sources by 2030. Their plan calls for the networking of 11.5 TW of electricity produced by renewable energy in the following manner: 50% from wind power, 20% by concentrating solar radiation, 20% from photovoltaic devices, 4% each from hydroelectric and geothermal sources, and 2% from wave surges and tides. Although the realization of such a plan in just 20 years is sheer utopia, nonetheless the proposal does provide an opportunity for interesting debates from which important conclusions could be reached and possible solutions discovered.

We have seen that there is no problem regarding the availability of renewable primary resources (Figure 70). But this good news does not in itself solve the problem of meeting future energy demands. We must fabricate devices capable of *converting* these endless (renewable) energy fluxes into useful energy, and we must find suitable sites at which to install such devices. Two resources, though not limitless, are therefore needed: raw materials and space.

As you can guess from the considerations we made above on the photovoltaics situation in Italy, the availability of landmass needed is not an insurmountable problem. Globally, the Jacobson-Delucchi plan would require less than 2% of the Earth's landmass. If we were to leverage sea coastal areas appropriately for wind power production, the required surface would drop to less than 1%.

Insofar as raw materials are concerned, the situation is much more critical. Wind power production will not be limited by the availability of steel and cement, nor will the silicon-based photovoltaics because of the endless abundance of silicon. However, the turbine engines contain a rare chemical element – neodymium (Nd) – whose current production is 100 times less than that which would be required by the Jacobson-Delucchi plan.

Another critical material is Lithium (Li), necessary to manufacture one million electric batteries to store the energy produced from intermittent sources. An annual production of 26 million electric cars (half of all cars sold worldwide in 2010) would necessitate 260 000 tons of lithium. In the absence of recycling, as is the case today, the world's reserves of this metal would be exhausted within 50 years. The present reserves of Li are mostly found in Bolivia, Chile, and Argentina – countries that would become strategically crucial in any future energy scenario, with the risk of creating South-American geopolitical problems similar to those that have characterized the Middle East during the era of crude oil.

The best catalysts for fuel cells (which, as seen in Chapter 7, are the key devices of a future hydrogen economy) contain Platinum (Pt), a very rare and expensive precious metal. Likewise, highly efficient electroluminescent devices use semiconductor elements (Indium, In; Gallium, Ga) or metals (Iridium, Ir), and alternative photovoltaic materials are based on Tellurium (Te) and Indium. All these elements are rare and expensive.

Of course, future scientific research could lead to the discovery of new materials and novel technologies to exploit renewable energy sources. It is in this field that we need to concentrate our collective resources and considerable intellectual capital. But sooner or later, we will have to face up to the physical limits of Spaceship Earth, from which we cannot escape, at least in the foreseeable future.

We have made a 180 degree turn back to the starting point. The endless flux of solar energy and human ingenuity alone will not save humanity from future energy, climate, and environmental crises.

The transition to different energy forms will require a radical change in the mindset of people, a change in their lifestyles, and a change in their entrenched practices. What is needed is a substantial injection of sobriety, good sense, foresight, and a good dose of responsibility. This applies to those countries, like the United States and Italy, which for decades have been living beyond their natural and economic resources, thus placing a heavy burden on the shoulders of future generations – our children and grandchildren.

15
Scenarios for the Future

> *A politician thinks about the upcoming elections; a statesman thinks about the next generation.*
>
> Alcide De Gasperi

When goods and services are produced it's as though we punched a hole in the Earth's crust to extract the necessary resources, generating at a later date a mountain of wastes (Figure 71).

Every action that mankind performs in the *technosphere* causes nature to be – more or less – depleted and contaminated. Therefore the continual increase in human activities must increasingly lead to a fear that the resources necessary to sustain life and the well-being of humanity may become irreversibly compromised – as in the case of the air we breathe – or exhausted – as in the case of fossil fuels.

Every barrel of oil burned today means a barrel of oil less for future generations and 320 kg of carbon dioxide discharged into the atmosphere. The problem of resource consumption and waste accumulation is therefore not only a problem of today's passengers on spaceship Earth, but even more so the problem of our great-grandchildren, who will have to deal with our legacy to them.

(Un)Sustainable Development

An idea has been taking hold in the last few decades that suggests we become fully aware of the physical limits of any growth, or better still that we pursue a *sustainable growth*. To a first approximation, this is defined as growth that *meets the needs of the present without compromising the ability of future generations to satisfy their own needs*. However, growth today is still understood as an increase in the production of goods and services, which cannot occur if we left an equivalent quantity and quality of resources to future generations. Such growth is simply not sustainable. If we continue along this route the maximum we can hope for is to pursue the course of a lesser unsustainable growth.

To achieve this goal, human activities should be evaluated on the basis of energy costs, on the cost of raw materials and on the environmental impact. With parity of economic value, we should look to goods and services that require less energy

Powering Planet Earth: Energy Solutions for the Future, First Edition. Nicola Armaroli, Vincenzo Balzani, and Nick Serpone.
© 2013 Wiley-VCH Verlag GmbH & Co. KGaA. Published 2013 by Wiley-VCH Verlag GmbH & Co. KGaA.

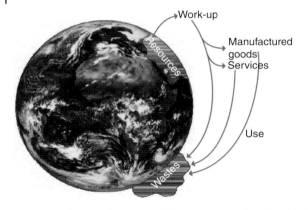

Figure 71   The use of non-renewable resources, such as fossil fuels, inevitably leads to the generation of wastes.

and fewer raw materials, that last longer, that produce less waste, and that involve less pollution and less consumption of natural resources.

America's Big Footprint

Various types of parameters can be used to quantify and discuss the problems of sustainability. The best known among these is the (ecological) *footprint*, defined as the area of the Earth's surface capable of providing the necessary resources for a person's daily consumption and for the disposal of wastes generated.

According to generally accepted estimates, the Earth can endure an average ecological footprint around 1.8 hectares per inhabitant (1 hectare = 10 000 m²). The latest estimates show that an average American citizen has an ecological footprint of 8 hectares, whereas for a Canadian it is 7 hectares, 5 for a German and Italian, 1.8 for a Colombian, 0.8 for an Indian, and 0.4 for an Afghan. Clearly, there are people who exploit much larger *slices of this Earth* than are rightfully theirs, while others use very small parts. Estimates indicate that the longevity of every United States citizen born today is, on average, 82 years and will use about 4 million kWh of electricity during this period, 200 million liters of water, 300 000 liters of fuel, and will produce 1600 tonnes of CO_2.

If each one of the 7 billion inhabitants presently on Earth had a footprint equal to that of the average American, we would need 4 Earth planets. The data suggest that not all the inhabitants of the planet can live as an American. Indeed, the day is fast approaching when even the North Americans will no longer be able to live as present-day Americans.

It is natural to think that rich countries should give a good example to reduce the unsustainable current growth. In fact, this doesn't happen because every call to consume less, particularly in the energy sector, contrasts with the notion backed by many economists, and adopted by most politicians, according to whom it is necessary that the gross domestic product (GDP) increase by at least 2 to 3% per year, even for rich countries. No wonder then that we are constantly pressured to consume more, to trade-in our cars every couple of years, and to discard other still useful household items just because some government incentive tells us so.

An increase in the GDP of developed countries will likely continue for several more years or even decades. In the meantime, such consumption will cause serious damage, whose remediation costs will have to be shouldered by the next generations.

The More We Consume, the More We're Happy?

Produce more! Consume more! Let the GDP grow! Is this really the recipe for happiness? Are we sure that today's farmers who produce more and faster are happier than the old farmer who planted small oak trees, knowing full well that only his grandchildren would reap the benefits of their shade?

Economic growth and well-being are spreading. Therefore, to measure our well-being, we should use indices that, alongside economic production, take into account social and environmental sustainability. For example, the GPI (*Genuine Progress Indicator*), which measures the increase in the quality of life of a country by distinguishing between the positives costs – such as those for goods and services – and the negative costs – such as those concerning crime, pollution, and traffic accidents.

The GPI and other similar indices have been proposed as alternatives to the GDP, which considers instead all expenses to be positive and does not include all those activities which, while not recorded as monetary expenditures, contribute to increasing the well-being of a society: for instance the non-remunerated work of housewives and volunteers. We see then that in developed countries, while the GDP continues to grow, the well-being of its citizenry seen through the GPI tends to decrease or remain flat, as illustrated for the USA in the years between 1980 and 2000 (see Figure 72).

A similar discussion applies particularly to energy consumption. We are led to believe that the quality of life always increases with increasing energy consumption. This may be true for the poorest countries, where energy consumption *per capita* is very low. However, when this consumption reaches approximately 2.6 toe per year (110 GJ/year, less than half the current average consumption in the Western world), a further increase does not lead to any appreciable improvement in the quality of life. In this regard, the infant mortality rate is slightly lower in Italy than in the United States, even though per capita energy consumption in the USA is more than double that of Italy. Numerous other indices confirm that the quality of life in developed countries does not increase with the consumption of energy. Rigorous analyses show that such countries could easily reduce their energy consumption by 30% without much sacrifice in their way of living – in fact

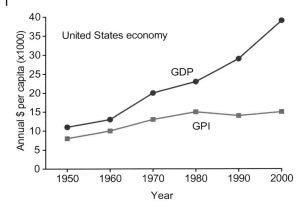

Figure 72 Comparison between the changes in GPI (genuine progress indicator) and GDP (gross domestic product) in the United States.

they could even draw some benefits, because excessive energy consumption – energy obesity – is very dangerous (see Chapter 3).

Dividing the current global primary energy supply (about 12 Gtoe, or 510 EJ) by the number of inhabitants of the planet (7 billion) gives approximately 75 GJ per capita, roughly equal to the amount of energy consumed by an average citizen of Western Europe around 1970, or by a citizen of the Balkans today. In other words, the quantity of energy currently consumed globally would allow all the inhabitants of the planet to enjoy a more than decent standard of living. Unfortunately, the reality is quite different. At present, Europeans consume three times more than they consumed in 1970 (180 GJ/year). Canadians and Americans would have to cut their consumption by about 80% to arrive at the equitable threshold of 75 GJ per capita. Certainly, it's not possible to force rich countries to return to the consumption levels of 40 years ago. However, the numbers cited above should at least give us some food for thought. We must ask ourselves: was life so difficult in 1970 in developed countries as to necessitate tripling energy consumption?

Nonetheless, other factors – for example, distances and weather conditions – must also be taken into consideration when attributing energy quotas to countries like Canada, where weather conditions are what they are: nearly 4–5 months of severe winter weather not to mention the long distances between cities, unlike continental Europe or the British Isles.

That's Enough!

More, greater, faster, hurry – are the orders we unconsciously obey in our daily consumption of goods and energy. Today, we use more energy, produce more cars, cut more trees, fish more fish, and so on. And yet, we forget that fossil fuels are non-regenerative resources, that a forest and fish keep to natural cycles and not economic ones, and that the atmosphere is not an infinite sink for the disposal of our toxic pollutants.

When we talk about the economy, consumption, and waste, it seems that words such as *too much* and *enough* ought to be censored or even completely eliminated. We seem to have lost any sense of limit or sense of proportion. Yet there are many signals out there, starting with the climate issue, that tell us we need to change course and move from an irrational increase in consumption to a sense of sobriety and sufficiency.

Collectively, we must come to understand that we can only go so far, but not beyond, and each of us must learn to say *enough is enough*. Saying *I have had enough* is to be in tune with nature and assert our social conscience. This is not a question of idealism; it's a question of necessity.

To be truthful, the concept of limit is already present in our individual consciousness. For example, even if I am nuts about chocolate, I know full well that if I were to eat 3 kg of it I would be very sick—so I try to limit myself. Unfortunately, this sense of limit is not yet part of our collective consciousness. Increased consumption of any kind is always considered a good thing. Politicians, economists and trade unionists repeat incessantly that we *must* increase consumption. One has to wonder which planet these people live on, and which planet they're talking about.

Our blind faith in an infinite *growth* is nothing less than a genuine defiance of the basic laws of nature, which finds its most glorious example in so-called *consumer credit*. Everything we do is within the reach of our pockets—for example, we can pay in convenient installments of 9.99 Euros ($15) a month or nothing at all for the first 12–18 months—as is often seen in North America. This is nothing but a ploy to encourage the consumer to buy at any cost—a sure way to personal and financial ruin. But who cares if the consumer faces financial hardships down the line? What's important is that the Gross Domestic Product continues to grow.

Strategies

The history of human civilization can be seen as the progressive growth of new energy resources and invention of new technologies to use such resources. Although it's true that knowledge and information are gaining increased importance with respect to the availability of raw materials, nevertheless the fact remains that energy availability determines, guides, restricts, and molds our ability to function in all societal undertakings.

In the current historical phase, we must realize that fossil fuels are a unique quantitatively limited wealth that Nature has been benevolent enough to provide for mankind. We are well aware that prolonged and indiscriminate use of this gift of Nature causes serious consequences to mankind and to the environment. Faced with these incontrovertible facts, we must adopt one of three possible strategies in our energy perspective, each one different from the other.

The *first strategy* would be to increase the indiscriminate use of fossil fuels, and to find other deposits in other regions of the Earth. This is the strategy of the ostrich, which buries its head in the sand because it doesn't want to see the error of its ways. Sooner rather than later, we will reach a point when fossil fuels are exhausted. But before that happens, we may have to face up to an irreversible ecological crisis that may arrive unexpectedly at any time. The clock is ticking!

The *second strategy* is not to rely on renewable energies because, as some maintain, they will fail to secure the enormous amount of energy required to sustain our current growth. To ensure that *growth* continues, this particular strategy relies on expanding the use of nuclear energy, which, as estimates indicate, may have little chance of success on the long term and be irrelevant in the short term.

The *third strategy* is based on the principle that we become aware of the physical limits of the Biosphere and we reduce our consumption of energy and raw materials. In this , we will have to rely heavily on a substantial expansion of renewable energies which, in the long run, will without doubt fulfill the majority of our energy requirements.

At the Crossroads

The energy question has placed society at the crossroads. On the one hand, there is the all-out defense of a lifestyle in rich countries based on high energy consumption – a lifestyle that damages the environment, that does not rule out violence or even war to seize the remaining reserves of fossil fuels and nuclear fuels wherever they may be, that doesn't take into account the rights of future generations, and that doesn't care to reduce the existing inequalities among the Earth's inhabitants, exposing itself to risks of nuclear proliferation.

On the other hand, the radical change in lifestyle imposed upon us by the necessity of physical constraints should also be an ethical choice – a lifestyle based on low energy consumption, on sobriety, and on sufficiency.

The second alternative requires a transitional period in which we reduce progressively the use of fossil fuels, avoid the expansion of nuclear power, and develop all types of widely available yet non-polluting renewable energy resources, each one valued according to the specificity of the region concerned.

In fact, if at present we don't know the details of what the future energy system will look like, we envisage each community having to take its responsibility on how to satisfy its own energy needs, taking into account the renewable resources available to that community. This approach is diametrically opposite to the current one, where energy supplies arrive from distant sites, often from other continents – a system that completely relieves the consumer of energy of any responsibility, deluding such consumer into thinking that energy supplies are infinite and that increasing production of wastes is justifiably OK.

Transition to Renewable Energy Resources

Transition toward renewable sources of energy will be a long venture from both technical and economic viewpoints. The cost of fossil fuels is still relatively low (in some countries), and the traditional energy infrastructures (oil wells, pipelines, gas pipelines, oil refineries) are widespread and well tested.

The transition will also be slow for social and political reasons. It will take time before Governments–with full knowledge of health care costs and economic damage caused by the use of fossil fuels–will intervene with appropriate policies and incentives and disincentives. In the meantime, we need to reduce wastes, increase efficiency in the use of energy, and launch extensive research and development programs on renewable energies.

The power density (Table 11) describes the amount of useful power that can be derived from a given source of energy for every square meter of the Earth's surface for its production. Table 11 also shows that the energy in fossil fuels is highly concentrated compared to diluted renewable energy forms. This is the element that determines the radical differences between the current energy system based on fossil fuels and the possible future scenario based on renewable sources.

The gradual shift of energy away from fossil fuels and the gradual transition to the more dilute renewable energies (that is, lower power density; see Table 13) will necessitate a substantial change in lifestyle. We'll have to get used to consuming less energy, particularly in the transport sector. In a sense, this will free us from being captives of the oil-producing countries–especially of the Middle East–because renewable energies are no longer localized in small areas of the planet, but are distributed more or less equally and are no longer owned by a handful of nations.

Table 11 Power per surface area obtained from various energy sources.

Source	Power density (W/m²)
Photovoltaic	20–60
Wind power	5–20
Hydroelectric (high altitude)	10–50
Hydroelectric (low altitude)	~1
Tides	10–50
Biomass	<1
Fossil fuels	1 000–10 000

Source: V. Smil, Energy at the Crossroads, MIT Press 2003.

The Scientist's and the Politician's Responsibility

It is time for scientists to speak frankly and publicly to people and, most importantly, to convince politicians that the exponential increase of energy consumption in many countries is a spiral that must be stopped, because it is untenable from the thermodynamic and ecological viewpoints, as well as being morally unacceptable. It is not possible to satisfy the continuous and unceasing increase in the energy demand. What is needed instead is a bold reduction in this demand. This is possible now, but does require the political will together with the understanding of the consumer.

Politicians must come to terms with the fact that energy is the big problem of the twenty-first century. It will be necessary to ensure an adequate energy supply to the 7 billion people–expected to increase to more than 8 billion in the next 20 years–to ensure an acceptable and peaceful coexistence among all of Earth's passengers without damaging the equilibrium of the biosphere.

Wise political choices in the field of energy are certainly not stimulated by articles and books that disseminate optimism about the health of the planet and the Earth's resources. To suggest, as some have done, that there is a great abundance of fossil fuels, that air pollution is not a problem, that global warming is a hoax, and that the dramatic increase in the world's population pose no problems is totally devoid of scientific basis. An even more preposterous claim is to suggest, as unfortunately some people are doing, that saving energy is not only useless but may even be harmful.

People need to be educated–more so the younger generation–beginning with the rational use of energy in schools, colleges, and institutes at all levels. Everyone must be made aware that a light left on, an electrical appliance left to operate or to be on standby, and driving a car (among others) not only come at a significant economic cost, but also at heavy environmental and social costs.

People must understand that we are on the threshold of a new era and that we must look far into the future. Development of the use of solar energy and other renewable energies is a look to the future–not nuclear energy, which would leave the immense burden of radioactive wastes to future generations. And we must also look far beyond our borders, because, unlike uranium and fossil fuels, solar energy and other renewable energies are abundantly available in all parts of spaceship Earth–they are free for the picking!

Challenges and Opportunities

At this point, you will likely think that we have painted a very depressing picture. Maybe we did, but as an ancient dictum said: *the only difference between an optimist and a pessimist is that the pessimist is better informed.*

We should recognize that it is precisely pessimism based on knowledge, and not optimism based on ignorance and misinformation, that will take us forward. Only if we know how things really stand and understand the issues will we be able to come up with solutions. For this to occur, however, requires – starting with the political class – that we all understand three fundamental notions: the first two are *inconvenient truths* for people in rich countries, while the third, though too often ignored, is a fact which gives mankind hope for a better future.

The two *uncomfortable facts* have been mentioned several times in this book:

1) The Earth's resources are limited, and therefore consumption cannot continue indefinitely.
2) Resources ought to be distributed more equitably among all people on this Earth.

These two hard realities must prompt us (i) to minimize, if not suppress, wastage altogether, (ii) to reduce consumption, (iii) to use resources more efficiently (first among which – energy), and (iv) to live with responsibility, solidarity, and compassion with each other.

The third undeniable reality is that spaceship Earth, in its journey throughout the universe, is always accompanied by the Sun, which supplies it with an inexhaustible amount of energy (fuel) – Figure 73. In an hour, Earth receives as much energy as that which we consume collectively in a year. The Sun will shine for billions of years to come. Its energy is non-polluting (in the traditional sense) and sustains life on Earth, and since it is diluted when it reaches the Earth it presents little danger to mankind.

Most importantly, solar energy is distributed well throughout the whole planet – a great additional benefit. The harnessing and use of solar energy and other renewable energies are therefore our challenges and great opportunities. If we learned to use them efficiently, they could solve not only the energy and ecological problems, but also the issue of inequalities among the Earth's people.

A final word to you, the reader – we do not claim that you'll remain impressed with the many things said and written in this book. That was not our goal.

Figure 73 Cartoon illustrating the Sun's gift to mankind. Source: http://solarenergydemystified.wordpress.com/2010/08/20/waking-up-to-solar/.

However, we would consider our job done if you remember at least one thing—no progress in the energy field will ever win over ignorance, waste, and the contempt for the limits of this biosphere **Earth** we like to call home.

The long and tiring journey of energy transition is not only a fascinating challenge at the scientific and technological level, but it is even more a cultural and moral challenge to take on individual responsibility. We are all asked to meet this challenge. And that includes you, too.

Time is pressing. The time to start is now!

Appendices

Appendix A: 17 Myths to be Dispelled

1. Nuclear energy is needed to ensure greater energy independence of Europe.
European countries do not have significant reserves of uranium. If they had, and assuming they produced all the electricity via the nuclear option, they would have met less than a quarter of Europe's energy needs toward end uses. More than three-quarters of Europe's current energy consumption is from fossil fuels – these fuels cannot be produced from nuclear power facilities.

2. We must turn to the nuclear option to counter the high price of oil.
Petroleum serves mainly to produce liquid fuels for transportation and is a feedstock for producing petrochemicals. Nuclear energy can only produce electricity. That the two things are completely disconnected is proved by the fact that with a very similar population density, France produces more than 75% of its electricity by nuclear power, but consumes more oil than Italy – Italy has no nuclear power facilities.

3. Italy needs to import nuclear electricity from France at high prices.
On the contrary, the need is not Italy's but France's. Nuclear power facilities cannot be switched on and off at will and operate continuously. During the night hours, when electricity demand is low, the French system technically needs to export its electricity surplus to neighboring countries, Italy included, so as to ensure the stability of the French system. The cost of this electricity is low, and thus welcomed by foreign power utilities.

4. To produce significant quantities of electricity by photovoltaics would require that all of Europe be covered by the photovoltaic panels.
With current technologies, far from being optimal, the surface of Europe to be covered to meet the electricity demand would be 0.6% of its landmass.

5. Biofuels will replace gasoline and diesel fuel.
Even with the use of bioethanol derived from sugar cane – the most efficient and cost-effective biofuel currently available from the energy point of view – to replace the 18 billion gallons of gasoline presently consumed in Italy would necessitate a huge and unrealistic area: about 35% of its farm land. If Europe and the United States wanted to replace even 5% of their consumption of fuels with biofuel

Powering Planet Earth: Energy Solutions for the Future, First Edition. Nicola Armaroli, Vincenzo Balzani, and Nick Serpone.
© 2013 Wiley-VCH Verlag GmbH & Co. KGaA. Published 2013 by Wiley-VCH Verlag GmbH & Co. KGaA.

products with current technology, they would have to dedicate about 20% of their arable land.

6. Today there are clean coal technologies.
Clean coal is a commercial and deceptive slogan. Coal remains the most polluting fossil source among those available today. The so-called clean coal involves confinement of CO_2 underground in caves but still produces large quantities of greenhouse gases, pollutants and ashes. This technology may take two or three decades to develop and most probably will never be available on a large scale and be economically competitive.

7. The new generation of fission and nuclear fusion reactors will shortly resolve the energy problem definitely.
The technical feasibility and cost-effectiveness of fourth-generation nuclear fission reactors and nuclear fusion reactors are yet to be proven. If these new technologies were to be implemented—doubtful according to some authoritative scientists—supporters of the nuclear option predict that this technology would be marketed only 30 to 40 years from now (2012). The world's energy problem cannot wait that long.

8. Solar energy can never meet the energy needs of humanity.
The endless flow of electromagnetic energy from the Sun is unique in quantitative terms, in that it can guarantee the fulfillment of man's energy needs in the long term. To solar energy we can add contributions from the thermal energy of the Earth's subsurface, also a source of immense potential.

9. Trading cars and appliances with government's incentives brings a benefit to the environment.
This statement may well be true at the local level, for example when new and less-polluting cars hit the road. But the balance at the global level is essentially negative in terms of consumption of energy and resources used to produce new goods and discard old ones.

10. The surface of the Earth floats on oil—it just needs to be explored and exploited.
One thing is to find small to medium size deposits—like those in the Basilicata region of Italy—but quite another to find supergiant deposits of good quality oil at low extraction costs in easily accessible areas, such as those of Saudi Arabia. Deposits of this type have not been found for decades and there is widespread skepticism about whether such deposits will ever be found again—all this in the face of the ever-increasing demand from the relentless economic growth of China, India, Brazil and Russia.

11. Methane does not pollute.
Methane is certainly the least polluting of fossil fuels, but does pollute nonetheless. Its combustion with oxygen in the air produces oxides of nitrogen, ultrafine particulate matter, and carbon dioxide.

12. Hydrogen is the clean energy of the future.

First of all, hydrogen is *not* a source of energy but is an energy carrier like electricity. To use it you must first produce it. This requires consuming energy. Extracting it from water requires at least the same amount of energy as that liberated when hydrogen reacts with oxygen to give water. We can't even claim that hydrogen is *clean*, since its cleanliness depends on the energy source used to produce it in the first place. Hydrogen will be the clean energy carrier of the future *only if* it is produced using renewable sources – for example, solar energy.

13. Rejection of the nuclear option puts Italy at a disadvantage compared to other industrial powers.

On the contrary, because it does no have nuclear fission, Italy is free from the *nuclear debt* that weighs heavily on those countries that have built and operate several nuclear reactors. Such countries, while living with a risk of a nuclear accident, in the future, will have to bear the high costs of decommissioning nuclear reactors and resolve the problems connected with the radioactive wastes.

14. Apart from very rare accidents, nuclear technology has proven to be safe and reliable.

To date, at least 6 reactors have undergone meltdown of the nuclear core, out of 594 worldwide built since the beginning of the civilian nuclear era – a disastrous performance for any technology. For example, suppose that 1 plane out of 100 were to crash during its lifetime because of structural failure, would you still accept flying?

15. Nuclear accidents are not so serious – even the Fukushima disaster, after all, has killed no one.

Chernobyl certainly did! And to a certain extent so did Fukushima. Unlike other types of accidents, a nuclear one is not definable in space or time. In fact, radioactivity – which causes serious illness – is transmitted largely through the atmosphere and the food chain, a process we cannot control. In addition, land use for agriculture is compromised for a very long time, maybe forever.

16. In the future we will need increasing amounts of energy.

This is certainly not true in developed countries. The directive 28/2009 of the European Union – so-called *20/20/20* – in essence *imposes* a 20% decrease in final energy consumption by 2020. An undisputed fact has now been translated into law. Today's technology allows us to live well – indeed, better, even if we consume less energy than before.

17. Renewable energy will never be sufficient to replace uranium and fossil fuels.

The potential of renewable energy that is technically exploitable and environmentally sustainable using today's technologies far outpaces current consumption. Solar energy is already exploitable, and can provide more than 20 times the current global primary energy consumption. In practice, the world can run with renewable energies, even though many of the technologies have not yet matured.

Appendix B: Maybe You Didn't Know That . . .

Consumption

Today, we consume worldwide *every second* 1000 barrels of oil (i.e., 159 000 liters), 100 000 cubic meters of natural gas, and 222 tons of coal.

The extraction cost of crude oil from Saudi Arabia deposits is $5–6 dollars per barrel, whereas the cost of extracting oil from a well deep in the ocean can reach $50–60 dollars per barrel.

It takes a lot of water to extract synthetic crude oil from oil sands. For example, it takes about 2–4.5 cubic meters of water to produce 1 cubic meter of synthetic crude oil.

The standard extraction process for the oil sands also requires large quantities of natural gas. At present, the oil sands industry uses about 4% of the Western Canada Sedimentary Basin natural gas – this is likely to increase to about 10% by 2015.

Today, the consumption of electricity in Italy is 20 times higher than it was in 1938.

An American citizen consumes as much energy as 2 Europeans, 4 Chinese, 14 Indians, and 240 Ethiopians.

In spite of being the third among oil producers and the first among the producers of natural gas and coal, the United States imports annually about 60% of oil and 6% of natural gas for its consumption.

To fabricate a personal computer requires an amount of energy equivalent to 250 kg of oil. Before being turned on, the computer has already consumed three times the energy it will use throughout its lifetime.

A mid-sized car consumes 1 liter of diesel fuel every 15 kilometers. With the same fuel, an Abrams army tank covers only a distance of 420 meters.

The energy needed to raise a 500-kg cow is equivalent to 6 barrels (about 1000 liters) of oil. To produce 1 kg of beef takes 7 liters of oil.

To fill an SUV gas tank with bioethanol necessitates a quantity of corn enough to feed a person for 1 year.

Powering Planet Earth: Energy Solutions for the Future, First Edition. Nicola Armaroli, Vincenzo Balzani, and Nick Serpone.
© 2013 Wiley-VCH Verlag GmbH & Co. KGaA. Published 2013 by Wiley-VCH Verlag GmbH & Co. KGaA.

Transportation

In 1901, public transportation in London, England, was provided by approximately 300 000 horses.

With 311 million inhabitants, the United States has 842 motor vehicles per 1000 people, including infants. China and India with a total population of 2.5 billion people have, respectively, 36 and 13 vehicles for every 1000 inhabitants. In practice, the number of vehicles in China today is at a level comparable to that of the United States around 1930.

Vehicles circulating in the United States consume roughly 5% of *all* world primary energy.

In Europe, car traffic is the cause of more than 30 000 casualties and over 1.5 millions wounded annually.

Nuclear Energy

From 1990 to 2010, the number of nuclear reactors in the world remained constant at 440 units. After the Fukushima incident, the number has decreased. Some Governments have decided to shut down their nuclear power facilities that have *suddenly* become obsolete and/or dangerous.

Plutonium-239 is so toxic that inhaling less than a millionth of a gram could cause lung cancer. From the beginning of the atomic era, nuclear electricity power stations have produced about 1500 tons of ^{239}Pu.

The United States storage area of radioactive wastes at Yucca Mountain was selected in 1987 after a 10-year selection process. It was supposed to come into operation by 2020. However, in 2009, the project was canceled for economic and security reasons, even though construction had reached an advanced stage.

If we wished to develop nuclear energy over the next 40 years so as to: (i) replace the current 430 reactors, (ii) replace half the current coal usage, and (iii) cover 50% of new demand for electricity, then we need to build approximately 2500 new 1000-MW nuclear reactors – that is, one per week from now to 2050. Clearly, this is an entirely unrealistic scenario.

The tsunami that hit Fukushima on March 11, 2011 had waves 14 meters high, more than double the protection wall of the nuclear power plant.

The tentative estimated damage following the accident of Fukushima runs between 100 and 200 billion Euros, equal to the cost of construction of 30–50 new nuclear reactors. For comparison, the compensation fund imposed by the US Government for the BP disaster in the Gulf of Mexico caused by the Deepwater Horizon platform was 20 billion dollars (ca. 14 billion Euros).

At the referendum of June 2011 in Italy, 57% of eligible voters voted – of these, 95% favored the repeal of the Government's decision to go back to the nuclear option.

Renewable Energy Sources

The Sun will continue to shine for billions of years, sending on Earth *every hour* 400 million billion joules of energy, equivalent to the energy that mankind consumes in an entire year. This gigantic energy flux is distributed more or less uniformly over inhabited areas of the planet. London (England) receives nearly two-thirds of the solar energy that Rome receives.

In 1979, President Carter had solar panels installed on the roof of the White House. In 1986, President Reagan had them removed.

At the end of 2011, there were worldwide installations of solar thermal panels that produced 232 GW, with an increase of ca. 20% compared to the previous year. In Austria, there are more than 512 m^2 of solar thermal panels per 1000 inhabitants; Italy has less than 34 m^2 per 1000 inhabitants.

Every year, Italian citizens spend nearly 150 billion Euros to drive their cars and buy 2 million new cars annually. Investing similar figures for the energy transition in 2030, Italy could run the country with 100% renewable energy, and could become a major exporter of energy.

At the end of 2011, the installed worldwide photovoltaic power amounted to 67.4 GW, an increase of about 68% compared to the previous year. Power *per capita* from photovoltaics in Germany is 3.5 times greater than that of Italy.

In July 2012, photovoltaic power installed in Italy was 14.6 GWp and produced 7.3% of the electric energy demand for that month.

At the end of 2011, the installed wind power in the world stood at over 238 GW (+21% compared to the previous year) and could cover almost 3% of the world's electricity demand.

The bottleneck of energy transition is not – as many argue – the availability of renewable energy, which is essentially endless, but that of raw materials for *fabricating* devices that convert renewable energy into useful energy.

Systems that concentrate solar energy are under construction worldwide for a total of 1 GW. It is expected that such facilities will produce a total power of 14 GW in 2014, to rise to 25 GW by 2020.

There are today 800 000 hydroelectric dams in operation worldwide, 45 000 of which are large dams higher than 15 meters.

Wastes and Pollution

Each year, Italians waste a quantity of food that could feed over 44 million people for an entire year, three-quarters of Italy's population. This waste has a strong environmental impact – improper waste disposal consumes 105 million cubic meters of water and produces 9.5 million tons of CO_2.

To produce a use-and-discard battery requires from 40 to 500 times more energy than that which it gives back.

Doing laundry at 90 °C consumes seven times more electricity than laundry done at 30 °C. A dryer consumes about twice the electricity used by the washing machine for the same load.

To produce 1 ton of aluminum takes about 16 000 kWh of electrical energy, equal to the consumption of a European household of four people for 5 years. To recycle 1 ton of aluminum requires only 800 kWh.

In the United States, only 44% of primary energy is transformed into useful energy: 56% is lost. Waste generated annually by 460 coal-fired electricity installations (with American coal) would fill more than three long trains from New York to Los Angeles.

More than 30% of CO_2 emissions of the rich countries of Europe is *imported* using goods produced from emerging economies, particularly China.

About 250 billion e-mails are sent every day in the world. An email of one megabyte (1 Mb) consumes an amount of energy equivalent to the production of approximately 20 grams of CO_2. For comparison, the Fiat 500 emits 119 grams of CO_2 per kilometer.

To take a hot shower consumes about $0.2 \, m^3$ of natural gas at a cost of 16 cents Euro. The same amount is spent watching television for 8 hours.

Disparity

Every minute, 1 Italian is born, 15 Chinese, and 37 Indians. Every day, the population of the planet grows at more than 200 000 units (the population of a medium-sized European city). Every year, there are about 80 million more people (equal to the sum of the populations of France and the Netherlands).

In the United States, about 28 million people (about 9% of the population) receive "food coupons" from the Government.

For some years, the World Food Day (October 16), which reminds us of the billion hungry people in the world, joins the World Obesity Day (October 10) to remind us that there are 1.5 billion obese people.

An American produces about 18 tons of CO_2 annually, nearly triple the amount produced by an Italian, who in turn produce five times more than an Indian.

Global average primary energy consumption *per capita* is about 75 GJ, equivalent to the national average consumption in Western Europe around 1970. Today, every European consumes nearly triple this amount, while an American consumes six times as much.

Appendix C

Table – Metric conversion factors.

Type of unit	United States unit	Equivalent in metric units	
Mass	1 short ton (2000 lbs)	0.907 184 7	metric tons (t)
	1 long ton	1.016 047	metric tons (t)
	1 pound (lbs)	0.453 592 37	kilograms (kg)
	1 pound uranium oxide (lb U_3O_8)	0.384 647	kilograms uranium (kg U)
	1 ounce avoirdupois (avdp oz)	28.349 52	grams (g)
Volume	1 barrel of oil (bbl)	0.158 987 3	cubic meters (m^3)
	1 cubic yard (yd^3)	0.764 555	cubic meters (m^3)
	1 cubic foot (ft^3)	0.028 316 85	cubic meters (m^3)
	1 U.S. gallon (gal)	3.785 412	liters (L)
	1 ounce, fluid (fl oz)	29.573 53	milliliters (mL)
	1 cubic inch (in^3)	16.387 06	milliliters (mL)
Length	1 mile (mi)	1.609 344	kilometers (km)
	1 yard (yd)	0.914 4	meters (m)
	1 foot (ft)	0.304 8	meters (m)
	1 inch (in)	2.54	centimeters (cm)
Area	1 acre	0.404 69	hectares (ha)
	1 square mile (mi^2)	2.589 988	square kilometers (km^2)
	1 square yard (yd^2)	0.836 127 4	square meters (m^2)
	1 square foot (ft^2)	0.092 903 04	square meters (m^2)
	1 square inch (in^2)	6.451 6	square centimeters (cm^2)
Energy	1 British thermal unit (Btu)	1 055.055 852	joules (J)
	1 calorie (cal)	4.186 8	joules (J)
	1 kilowatt-hour (kWh)	3.6	megajoules (MJ)
Temperature	32 degrees Fahrenheit (°F)	0	degrees Celsius (°C)
	212 degrees Fahrenheit (°F)	100	degrees Celsius (°C)

Source: U.S. Energy Information Administration / Annual Energy Review 2010.

Powering Planet Earth: Energy Solutions for the Future, First Edition. Nicola Armaroli, Vincenzo Balzani, and Nick Serpone.
© 2013 Wiley-VCH Verlag GmbH & Co. KGaA. Published 2013 by Wiley-VCH Verlag GmbH & Co. KGaA.

Appendix D: Bibliography

In this book, we have quoted a lot of data, and you – the reader – may rightly wonder where we got it from. Energy is a **hot** current theme debated daily in the mass media, which too often flood its readership with contradictory and confusing information.

The media never reveal their sources of information and don't feel obliged to verify the trustworthiness of the information printed or otherwise broadcast – typically, the more sensational the news, the better. Unfortunately, the rate of sensationalism is inversely proportional to the reliability of the news. If you see newspaper headlines such as – *Earth's climate is cooling; Water-driven engine invented; Nuclear wastes to end up in outer space* – then we suggest you turn the page and forget about such ridiculous *sensational* articles. **BUT** these headlines will sell papers and fascinate radio listeners and TV viewers.

Results of scientific research are **not** published in newspapers but in scientific journals, only after they have undergone a strict **peer review** process involving two to five experts (scientists acting as *referees*) who perform their evaluation in complete anonymity.

Scientists review each other's work, which places severe limits on the possibility of misunderstandings and falsification of data. Journal publishers and editors alone decide whether to publish the article on the basis of reviewers' (referees) evaluations. Scientific and technical journals are classified on the basis of their reputation and on their readership, while researchers are evaluated on the basis of the number of publications and number of citations of their articles in other publications. The reader can easily verify the scientific production of any researchers by typing their names in the Google Scholar search engine http://www.scholar.google.com.

Researchers are classified worldwide in a manner that vaguely resembles that of tennis players. Though the system may not be perfect, it does aim to promote quality and merit. (See, for example, the ranking of the most cited Italian scientists in the international literature compiled by the VIA-Academy on its website http://www.via-academy.org.). Unfortunately, these criteria are almost unknown to the media and to the public at large. Often the opinions of knowledgeable and authoritative scientists appear on the same playing field as those of less-qualified colleagues. This tends to short-circuit the message, misinform, and even confuse interested readers.

Powering Planet Earth: Energy Solutions for the Future, First Edition. Nicola Armaroli, Vincenzo Balzani, and Nick Serpone.
© 2013 Wiley-VCH Verlag GmbH & Co. KGaA. Published 2013 by Wiley-VCH Verlag GmbH & Co. KGaA.

To the extent that this book was intended mostly for the general public, we have refrained from giving a complete list of the hundreds of references we have consulted. Nonetheless, we provide some website sources we have used, pointing out that the data were drawn from highly reputable international scientific journals, such as, for example, *Science, Nature, Proceedings of the National Academy of Sciences (USA)* and others, and from books authored by Vaclav Smil of the University of Manitoba, Canada.

With regard to the general data on energy production, energy consumption, and energy reserves, the main sources of information were reports and databases from the United States Department of Energy, the European Commission, the International Atomic Energy Agency (IAEA), British Petroleum, ENI and utilities of electricity and natural gas, among others.

The data suggested for Italy's *roadmap* towards 2020 are from the Commission of the Accademia Nazionale dei Lincei.

Sources of information for Energy Canada, Energy USA and Energy UK are from various websites that were accessible in December 2011.

All sources are public and therefore can be consulted.

Useful Websites

The number of Internet sites that discuss energy, resources, and environment is overwhelming. Herein we only report the most authoritative and comprehensive websites – most are in English.

General Energy Databases

www.eia.gov
Official statistical agency of the Government of the United States with recent and past data on all sources, even broken down by geographical areas and countries: consumption, prices, and projections. Very effective is the section for visitors.

www.energy.eu
In this European Union portal you can compare the price of fuels, natural gas, and electricity of all European countries, and even incentives on renewable energies. You'll discover that what you read or hear on the media is not always true.

http://www.worldenergy.org/publications/default.asp
Reports and statistics of the World Energy Council, an association of public and private entities that operate in the energy sector in over 100 countries around the world. Very comprehensive are the 3-year reports on all energy sources.

www.worldenergyoutlook.org
Statistics of the International Energy Agency (IEA), only a portion of which are available free of charge.

http://www.eea.europa.eu/themes/energy
Energy site of the European Environmental Agency (EEA) with numerous reports on all energy matters and their related environmental problems. All the reports are downloadable free of charge.

www.bp.com and www.eni.com
Annual reports on resources and consumption of fossil fuels from two major oil companies; they are considered among the most influential worldwide.

www.terna.it and www.autorita.energia.it
Data, statistics, and reports on electricity and gas in Italy.

Data on Resources, Pollution, and the State of the Planet

www.earthtrends.wri.org
The World Resources Institute is the most authoritative Association in the world in the environmental sector. Here you will find reports, maps, and statistics. A fantastic source!

http://www.wri.org/project/earthtrends/
United Nations Development Program: a huge amount of information.

www.footprintnetwork.org
The Global Footprint Network: data and discussions on the ecological footprint and related topics.

Renewable Energies

www.ren21.net
REN 21 is the International Forum on renewable energies. Here you will find numerous authoritative reports downloadable free of charge.

www.epia.org and www.ewea.org
European associations of photovoltaic energy and wind energy.

www.nrel.gov
National Renewable Energy Laboratory of the United States located in Golden, Colorado.

www.lincei.it
This website offers an Italian 2020 roadmap for renewable energies.

www.energystrategy.it
Website of the Polytechnic Institute of Milan–analyzes the strategies and technological choices in the field of renewable energies.

Nuclear Energy

http://www.nrc.gov/waste.html
Website of the United States Nuclear Regulatory Commission devoted to nuclear waste.

www.gen-4.org
International Forum on the so-called fourth-generation nuclear reactors.

www.iter.org
International project for controlled thermonuclear fusion (ITER).

Efficiency and Energy Education

www.eere.energy.gov
Energy efficiency and renewable energies from the United States Department of Energy.

http://www.co2now.org
Scientific information about the levels of concentration of CO_2 in the atmosphere and related risks.

Climate Changes

www.ipcc.ch
IPCC, the International Panel of the United Nations on climate change, received the Nobel Prize for Peace in 2007.

www.realclimate.org
Web forum where some authoritative international climatologists dismantle the increasingly discredited theses from skeptics on global warming.

For Children and Teachers

http://www.epa.gov/students/index.html
Environmental Protection Agency (EPA) of the United States; delightful accounts for students.

Index

Powering Planet Earth: Energy Solutions for the Future, First Edition. Nicola Armaroli, Vincenzo Balzani, and
Nick Serpone.
© 2013 Wiley-VCH Verlag GmbH & Co. KGaA. Published 2013 by Wiley-VCH Verlag GmbH & Co. KGaA.